# Electromagnetic Force

Developed at
The Lawrence Hall of Science,
University of California, Berkeley
Published and distributed by
Delta Education,
a member of the School Specialty Family

1465670
978-1-62571-175-5
Printing 1 — 9/2016
Quad/Graphics, Versailles, KY

# Table of Contents

## Readings

### Investigation 1: What Is Force?

The Force Is with You. . . . . . . . . . . . . . . .3
The Discovery of Friction . . . . . . . . . . . .8
Net Force . . . . . . . . . . . . . . . . . . . . . .15

### Investigation 2: The Force of Magnetism

Magnetic Force. . . . . . . . . . . . . . . . . . .19

### Investigation 3: Electromagnetism

Circuitry and Lightbulbs . . . . . . . . . . . . .25
What Is Electricity? . . . . . . . . . . . . . . . .31
Electromagnetism . . . . . . . . . . . . . . . . .38
Electromagnetic Engineering. . . . . . . . . . .42

### Investigation 4: Energy Transfer

The Rebirth of Electric Cars. . . . . . . . . . . .47
Where We Get Energy . . . . . . . . . . . . . . .56

## Images and Data . . . . . . . . . . . .63

## References

Science Practices. . . . . . . . . . . . . . . . . .73
Engineering Practices . . . . . . . . . . . . . . .74
Science Safety Rules . . . . . . . . . . . . . . . .75
Glossary . . . . . . . . . . . . . . . . . . . . . . .76
Index . . . . . . . . . . . . . . . . . . . . . . . . .78

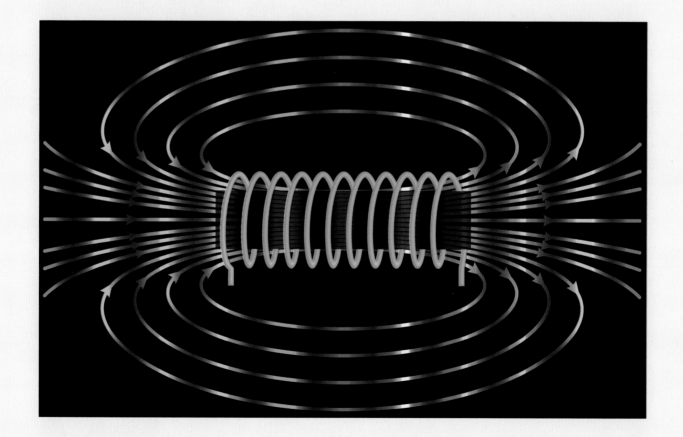

A skateboarder's push is one kind of force that produces motion. What other pushes and pulls have you experienced today? What moved because of the applied force?

# The Force Is with You

The skater stands at the edge of the half pipe, ready to push off. That first push is the force that sets her and the skateboard in motion.

If a skateboard is sitting still on a level sidewalk, it cannot move by itself. The skateboard will move only when a **force** acts upon it. Pushes and pulls are forces. Forces make things move.

**Pushes and Pulls**

Force

Motion

Force

Motion

As the skateboarder pushes and pulls her board, the red arrows show force and the blue arrows show motion. In both cases, the board moves in the same direction as the force.

## Force: An Interaction

In order to start moving, the girl must apply a force. A push is a force. If she wants her skateboard to go faster, she could apply additional force with another push. But no push exists without something doing the pushing and something to push on. A force is an **interaction** between objects. A force can happen only when one object pushes or pulls on another object.

Which objects interact to get the skateboard moving? The girl pushes on the ground and also pushes the skateboard forward. The girl and skateboard form one part of the system because they move together. When the girl pushes against the ground with one foot, the girl and skateboard start to move in the direction she applies the force. Imagine if there were no ground for the girl to push against. Without an object to push against, there would be no force to drive her forward.

What if the skateboard starts rolling too fast? How can the skateboarder stop it? If she applies a force opposite the direction of the motion, she can slow or stop the board.

## Starting and Stopping

Motion  Force

Existing motion

New force

**Skateboard stops moving**

When the skateboarder applies a force (red arrows) against the ground, she can change the motion (blue arrows) of the board.

With both feet on her board, this skateboarder is not pushing off the ground. But an invisible force will help her down the ramp. What is that force?

## An Invisible Force

When you let go of the skateboard at the top of a slope, it begins to roll downhill. Nobody is pushing or pulling it, yet it is moving. The invisible force being applied to the skateboard is **gravity**.

Gravity is a force of attraction between any two objects. The more massive the object, the greater the gravitational force. The skateboard is one object. Earth is the other object. The planet is massive, so it pulls strongly on all other objects, including the skateboard. It pulls the skateboard toward the center of Earth and down the slope.

When the skateboard is sitting on flat ground, you might not notice the force of gravity, but it is still acting on the board. Two forces are working on the skateboard. Gravity is pulling down on the skateboard, and the ground is pushing up on it. These two forces acting on the skateboard are equal, but in opposite directions. That means the forces are balanced. When forces are balanced, an object's motion does not change.

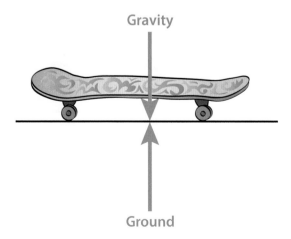

Gravity

Ground

## Measuring Force

A common unit of force is the **newton (N)**, named after Sir Isaac Newton (1643–1727). A **spring scale** is a simple piece of technology designed to measure force. The spring is made of tough steel wire wound in a coil. The spring is mounted in a clear cylinder. When you pull down on the hook at the bottom or push down on the top of the central **shaft**, the spring **compresses**. When you add a load to the hook, the force of gravity acting on the load's mass pulls the spring down. The downward force of gravity on a mass is its **weight**. Adding mass, which increases the weight, pulls the hook down further.

### Did You Know?

Newton was one of the greatest scientific minds of all time. He experimented with objects in motion. Based on his experiments, Newton used math to describe the movement of objects within the solar system, including comets and Earth's tides.

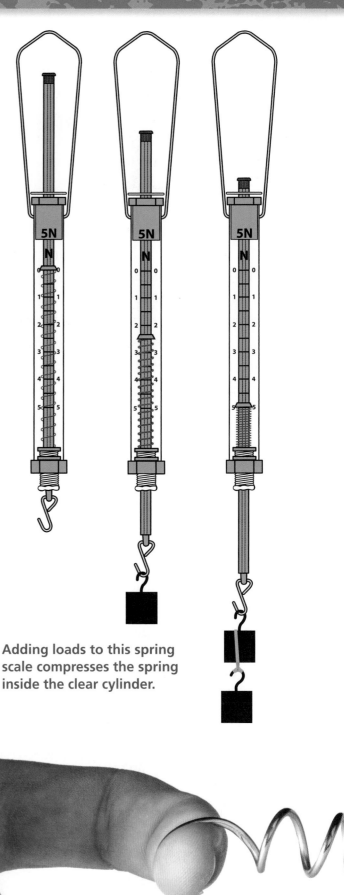

Adding loads to this spring scale compresses the spring inside the clear cylinder.

If you compress a spring in your hand, you can feel the force of the metal pushing back toward its original position. As the weight of the load compresses the scale's spring, the spring exerts more force upward. When the spring exerts an upward force equal to the load's weight, the hook does not drop any lower. The force of gravity pulling the mass down equals the force of the spring pulling the mass up.

A 100-gram (g) mass exerts a force of 1 N when pulled by the force of gravity on the surface of Earth. The spring scale is marked in 10 g intervals on one side for measuring mass and 0.1 N intervals on the other side for measuring force.

## Think Questions

1. What are the objects interacting in these actions?
   - A bicycle moves.
   - You throw a ball straight up in the air.
2. Look at the diagram with the spring scales. What is the force in newtons on each load?
3. What is the difference between mass and weight?
4. Do you know of any other invisible forces besides gravity?

A spring is a spiral of metal that squeezes together or stretches apart when you push or pull on it. Because of its elasticity, it goes back to its original length when you remove the force.

A 170-kilometers per hour slap shot might result in a winning goal, with the puck stopping when it slams into the net. The force that stops most rolling or sliding objects is less dramatic, and less obvious.

# The Discovery of Friction

A hockey puck slides across the slippery ice at high speed. In seconds, a player's hockey stick connects with the puck, applying a strong force to change its direction. But if no player hit the puck, would it have kept going in the same direction forever?

The answer has to do with forces, including **friction** and gravity. How do you measure a force you cannot see? How do you even know it is there? How did scientists come to understand force, friction, and motion? It took many inspired people a very long time to get it right, and it started with a bunch of wrong ideas.

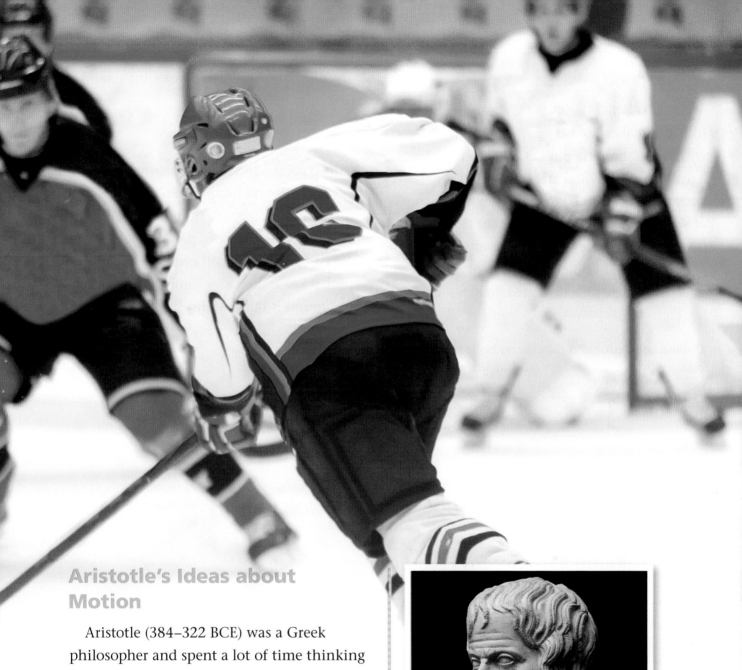

## Aristotle's Ideas about Motion

Aristotle (384–322 BCE) was a Greek philosopher and spent a lot of time thinking about all kinds of things. He thought and wrote about all aspects of the natural world.

Aristotle's ideas about motion were based on simple observations. He thought that some "natural" motions, like falling, seek a natural condition of rest. He thought other "forced" motions, like someone pulling a cart, continue as long as the force is applied. When the force stops, motion stops.

Aristotle was one of the greatest thinkers and teachers of all time, but his explanations of motion were incomplete. It is difficult to gather evidence about invisible forces. Many

**Aristotle's ideas helped people for centuries to understand how the world works. Later, scientists worked to expand on or correct these ideas.**

of Aristotle's incorrect explanations were accepted for almost 2,000 years. Explanations for motion changed in the late 1500s.

## Galileo's Physics

Galileo Galilei (1564–1642) was trained in mathematics. He became fascinated with understanding physics. Galileo tested Aristotle's ideas with careful experiments.

Galileo contributed to many fields of science, including the physics of the pendulum, ballistics, and astronomy. Many people consider Galileo the father of modern science because he relied on systematic observation, measurement, experiment, and testing hypotheses.

## Friction Experiments

Galileo conducted many experiments involving balls and ramps. He realized that all changes of motion result from forces. Objects fall because of the force of gravity. He also observed that when a ball comes to the end of a ramp, it keeps going for some distance, even if the ground is flat. But just like the hockey

**Galileo used grooved ramps and balls to watch what happened as a ball rolled down a hill and onto a flat surface. His explanations revolutionized our ideas about how objects move.**

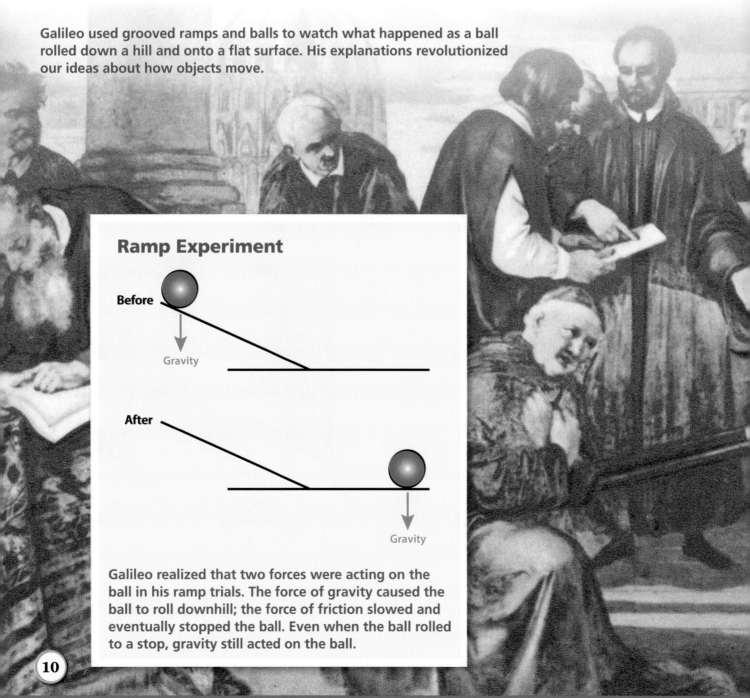

### Ramp Experiment

Before

Gravity

After

Gravity

**Galileo realized that two forces were acting on the ball in his ramp trials. The force of gravity caused the ball to roll downhill; the force of friction slowed and eventually stopped the ball. Even when the ball rolled to a stop, gravity still acted on the ball.**

puck in the ice rink, the ball eventually slows to a stop. Reversing Aristotle's notion, Galileo hypothesized that moving objects naturally stay in motion, and another force acts to change the object's motion.

Friction is a force between surfaces. The force of friction eventually slows the puck to a stop. The surface of the ice creates less friction than a rough sidewalk or grassy field, but friction will eventually stop the puck. Galileo realized that the force of friction is extremely important in the way things move on Earth.

This important discovery was theoretical. It was impossible for Galileo to do an experiment with no friction. He did not have computer simulations to help him test motion without friction. He developed mental and mathematical models of the world. Then he described his thinking for others.

## Apple Falling from Tree

Newton's falling apple observations led to a historic discovery: Gravity is a force that acts between all objects that have mass. It affects the motion of everything on Earth and in space.

## Newton's Physics

The year that Galileo died, Sir Isaac Newton (1643–1727) was born in England. He liked to read and think alone, often lost in thought. He was an excellent student and studied mathematics at Cambridge University.

At Cambridge, he designed equations that would become the foundations of calculus. When the university closed because of the black plague, Newton went home to develop his ideas.

At the age of 22, Newton had his now-famous idea about gravity. When an apple fell from a tree nearby, Newton realized that the apple fell down because Earth pulled on it. At the same time, the apple pulled on Earth. Newton was the first to explain the force of gravity as an attraction between two masses.

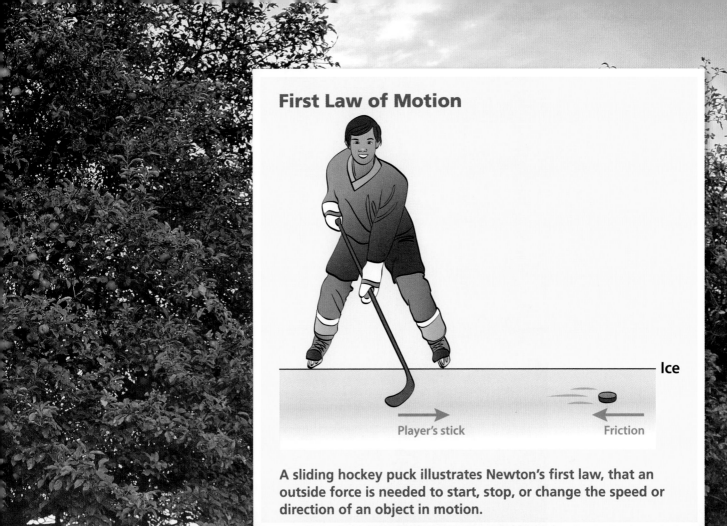

**First Law of Motion**

Ice

Player's stick

Friction

A sliding hockey puck illustrates Newton's first law, that an outside force is needed to start, stop, or change the speed or direction of an object in motion.

## Laws of Motion

Newton was aware of Galileo's work. He spent many years expanding on Galileo's ideas about force and motion. In 1687, Newton used mathematical evidence to explain his laws of motion.

1. Every object stays at rest unless an outside force acts on it. If it is moving, it travels in a straight line unless an outside force acts on it.
2. The **acceleration** of an object is directly proportional to the force on the object. It is inversely proportional to its mass.
3. For every action (force), there is an equal and opposite reaction (force).

These three laws explain the physics of motion. You have already learned about the first one in this course.

Let's apply Newton's first law to that hockey puck sliding on the ice. In motion in a straight line, it is sliding away from the players. Just as Galileo realized, the outside force acting on the puck is friction. Friction is slowing it down.

In addition to developing the laws of motion and universal gravitation, Newton investigated the nature of light, using prisms to discover that white light is a mixture of colors.

## Collaboration

In science, people collaborate and are inspired by what others have discovered. Newton's ideas were influenced by the work of Aristotle, Galileo, and many others. He said this about his great discoveries: "If I have seen further [than most people], it is by standing on the shoulders of giants."

Some people consider Newton to be the greatest scientist of all time. His great leaps forward in explaining the natural world were at first difficult to accept and understand. But they inspired other scientists and led to many more collaborations over the centuries.

### Think Questions

1. How does the texture of a surface affect friction?

2. What might happen if you tried to walk across the room and there was no friction?

3. What do you think Newton meant when he said, "If I have seen further [than most people], it is by standing on the shoulders of giants"?

# Net Force

Tug of war is a fierce game. Each team holds the rope tightly, waiting for the signal. The whistle blows, and the competition begins. Which team will win? It all depends on the force each team applies.

You know that a force is a push or a pull. You also know that an object will stay at rest unless a force that is strong enough acts on the object. In tug of war, opposing teams apply opposing forces through a rope. If each player in the picture could pull with exactly the same force, the game would be a tie. Neither team would move the other team.

**As long as each team is pulling with an equal amount of force, the forces are balanced and the rope does not move. What will happen if someone joins the team on the left?**

A seesaw is not much fun unless someone sits on the other end. Then, depending on the net force, this girl will go up or down. It takes two to seesaw!

Let's assume each player pulls with 200 newtons (N), so the total force one team is exerting would be

(+200 N) + (+200 N) + (+200 N) = +600 N.

Since the other team is pulling in the opposite direction, each one has a negative force. Their total force is

(−200 N) + (−200 N) + (−200 N) = −600 N.

The **net force** on the rope is

(+600 N) + (−600 N) = 0 N.

If one player pulls with 201 N, the total force of the team would become 601 N. Their team would win unless the opposing team members could meet or exceed the extra force.

Every day, you encounter situations that involve net force. Net force determines whether an object moves. Think about the forces involved in these examples.

- Different-sized people sit on a seesaw and push off the ground.
- You stand on a moving bus and try to not to fall over when the driver hits the brakes.
- You lean back in a chair without tipping over.
- You sit in your chair without floating off into space.

## Finding Net Force

Some objects simply lay on a table. Some people might think that no forces are acting upon the objects. But you know that is not true. Any net force other than zero will make an object move. As in the tug-of-war analysis, a little math helps us figure out the net force on an object.

What are the forces on the laptop? Because the laptop is not moving, we know the forces acting on it are balanced. Two equal forces act on the laptop in opposite directions. Gravity pulls the laptop down, and the table pushes the laptop up. All the forces acting on the laptop together represent the net force.

We know the net force on the laptop is zero, because it is not moving. If gravity is pulling the laptop down with a force of 20 N, what is the force of the table pushing up? If you said 20 N, you are correct. Equal forces are acting on the laptop in opposite directions.

Now consider the pen on the table. The pen has a smaller mass than the laptop. The force of gravity depends on the mass of the object, so the force of gravity pulling down on the pen is just 0.4 N. What is the force of the table pushing up? We know that it must be 0.4 N.

**Explain what forces are represented by the up and down arrows.**

0.4 N
0.4 N
20 N
20 N

Balanced forces keep the laptop and the other objects in place on the tabletop instead of falling down or flying up.

Some forces are easy to see, like a kick to a soccer ball. Other forces are invisible, like gravity, friction, and wind. Can you see the effects of these unseen forces acting in this room?

## Invisible Forces

How does the table "know" to push up on the laptop with a force of 20 N and on the pen with a force of 0.4 N? The difference in force depends on invisible properties of matter.

A gentle breeze comes from the window. Papers on the desk move to the side, fall, and come to rest on the floor. What happened? The force of the breeze (0.2 N) was greater than the force of friction on the paper (0.1 N). Even a tiny net force will cause motion. During all these motions, the net forces on the papers were not zero, so the paper fell to the floor. When the paper comes to rest on the floor, the net forces are again zero.

As you continue to study force, you will encounter other forces that are invisible. But their effects can be directly observed. What causes a sock to stick to your shirt when you take it out of the dryer? What causes a **magnet** to hold your report card to the refrigerator? What causes your bedroom light to turn on? The short answer is force.

### Think Questions

1. Choose one example from the bulleted list on page 16. Draw a model of how forces are acting upon the object (blue arrows) and how they affect the object's motion (red arrows). Use larger and smaller arrows to represent the strength of the forces.

2. Consider the laptop on a table. Let's say you apply a little horizontal push (force) to the laptop and it does not move sideways. What force is resisting the laptop's motion? Draw a diagram to explain your answer.

3. Think about tug of war. Team X has three members who pull with the following forces: +180 N, +200 N, and +190 N. Team Y has three members who pull with the following forces: −200 N, −180 N, and −191 N. Which team would win?

# Magnetic Force

The sky is overcast with thick clouds. A storm is coming in. You and your friends have been trying to find your way back to the ranger station for an hour.

Whose idea was it to go off the trail? That clump of rocks looks awfully familiar. You're pretty sure you are going in circles. If only you had brought a **compass** to show which way is north!

## Fields of Force

When you are using a compass, the tiny magnet in the compass needle aligns with Earth's **magnetism** and helps you find your way. But closed in its plastic case, the needle never touches another magnet. How do magnets exert a force without touching?

**Earth is a giant magnet with poles and a magnetic field that reaches far into space. A compass needle points north when it interacts with this planet-sized magnet.**

It is similar to how a falling apple and Earth exert a force of gravity on each other. Both the apple and Earth have a gravitational force that extends from them because of their masses. It forms an invisible **gravitational field**. Like gravity, magnetism is another invisible force of nature. Magnetism extends out from a magnet into the surrounding space to form what is called a **magnetic field**.

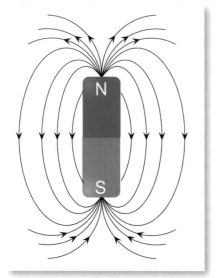

A field of magnetism extends out from every magnet. The magnetic force is strongest near the magnet's north and south poles.

## Force of Attraction or Repulsion

You feel magnetic force when you try to separate two magnets that are stuck together. You also feel magnetic force when you push two magnets together and they push away from each other. Magnetic force makes magnets act the ways they do.

The magnets used in class are **permanent magnets**. They exhibit magnetic properties pretty much all the time. Every magnet has two different sides or ends called **poles**, the north pole and the south pole. A simple bar magnet has its two poles on opposite ends. A horseshoe magnet has a pole on each end of the horseshoe. The doughnut magnets you worked with have poles on the two flat sides.

What happens when you hold two magnets close to each other? They exert a force on each other, but will they **attract** or **repel**? It all depends on how the poles are oriented. Below are four pairs of bar magnets being held together. Which ones will push apart when they are released?

## Magnets Held Together

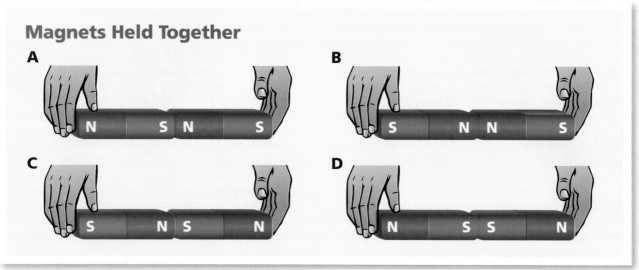

These pairs of magnets are held together in different configurations. What will happen when they are released?

## Magnets Are Released

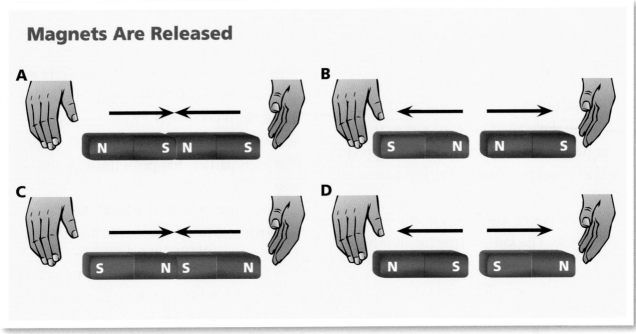

Observe how the magnets move when they are not held together. Opposite poles pull together, or attract, and like poles push apart, or repel.

The diagram above shows what happens when the magnets are released. Two general rules apply here. Can you figure out what the rules are?

The two pairs of magnets on the left attract each other. The two pairs of magnets on the right repel each other. Two north poles always repel each other. Two south poles always repel each other. We can state a general rule: like poles repel.

A north pole and a south pole always attract each other. It does not matter which magnet has the north pole and which has the south pole. We can state another general rule: opposite poles attract.

## How Magnets Stick to Iron

If opposite poles attract, why does a magnet stick to a piece of iron, like an iron nail, that is not a permanent magnet? Remember that magnetism extends out from a magnet in an invisible area called a magnetic field. When a magnet comes close to a piece of iron, such as an iron nail, the magnetic field interacts with the iron in the nail. The nail becomes a **temporary magnet**. The end of the nail becomes one pole of a magnet. The magnet then sticks to the temporary magnet.

So magnets do not really stick to iron. Magnets stick to other magnets, even if they are temporary. The temporary magnetism in the iron is called **induced magnetism**. Induced magnetism happens only when a magnet is nearby.

### Take Note

What are some examples of induced magnetism you observed in class?

## Particle Properties

To understand why some materials have induced magnetism and others do not, we have to explore the properties of magnets at the **particle** level. That means at the level of atoms and molecules. We can start to think about particles by considering what happens when a bar magnet breaks. Do you have a magnet with just one pole?

No, both pieces still have a north pole and a south pole. The same is true for all other magnets. No matter how many pieces you cut a magnet into, each piece still has a north pole and a south pole.

Each magnet piece has poles lined up the same way. If you did this a million million times, until you had the tiniest particle of the magnet that was still a magnet, you would see that each particle has poles lined up the same way.

This property defines a permanent magnet. Each particle has properties of magnetism. As

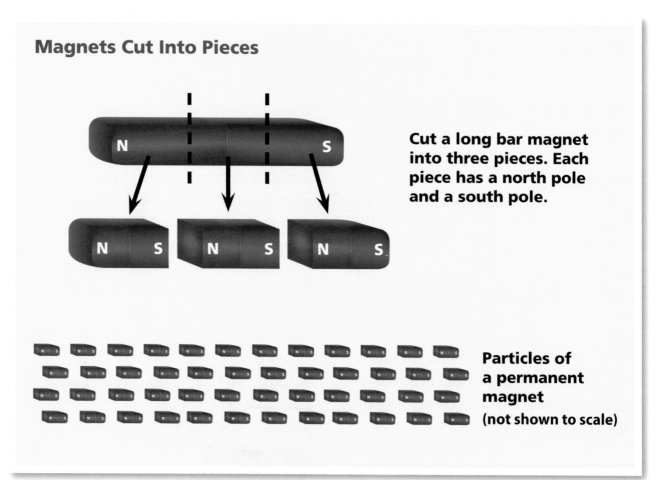

**Magnets Cut Into Pieces**

N · · · S

**Cut a long bar magnet into three pieces. Each piece has a north pole and a south pole.**

N S · N S · N S

**Particles of a permanent magnet**
**(not shown to scale)**

Even at the particle level, a magnet is still a magnet. Each atom of a magnet has a magnetic field and poles lined up in the same orientation.

the particles line up, each tiny magnetic field adds itself to form one big magnetic field.

## Nonmagnetic Materials

All nonmagnets can be split into two general categories, magnetic materials and nonmagnetic materials. Magnetic materials, such as the elements iron, nickel, and cobalt, have magnetic properties at the particle level.

But the particles are not all aligned the same way. Those particles can line up when they are in a magnetic field. The materials become a temporary (induced) magnet.

Nonmagnetic materials, like plastic, do not have magnetic properties at the particle level. Those particles are not affected when they are in a magnetic field.

**Materials outside a Strong Magnetic Field**

**Particles of iron**

**Particles of plastic**

**Magnetic at particle level**

(not shown to scale)

**Not magnetic at particle level**

(not shown to scale)

**The iron and plastic particles are oriented in different directions.**

**Materials inside a Strong Magnetic Field**

**Particles of iron**

**Particles of plastic**

**Temporary (induced) magnet**

(not shown to scale)

**Not magnetic**

(not shown to scale)

**The iron particles orient to the magnetic field and form a temporary magnet. The plastic particles do not change.**

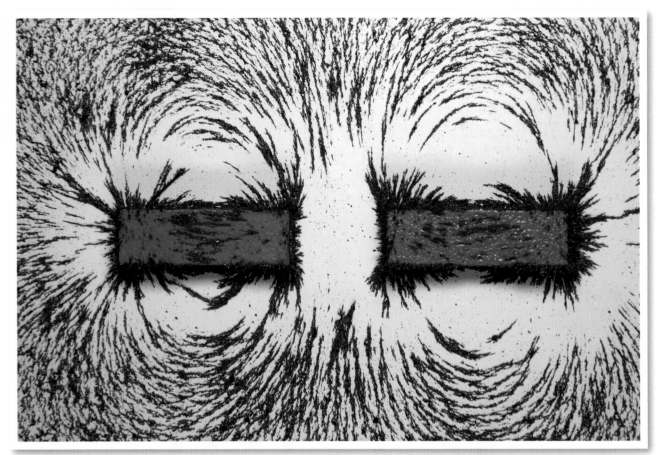

Iron filings spread around a magnet will form a pattern that shows the shape of the magnet's magnetic field.

If you bring a strong magnetic field close to a magnetic material, the particles in the material will line up with the magnetic field. They create weak temporary magnets. Particles in the nonmagnetic materials will not line up. So even the strongest magnet cannot attract or repel a material like plastic or wood.

1. **What rules determine whether magnets will attract or repel each other?**

2. **How can a magnet attract or repel another magnet even if they are not touching?**

3. **If you bring the south pole of a magnet close to the head of an iron nail, what changes will happen to the iron particles?**

# Circuitry and Lightbulbs

Energy makes things happen. There are many ways to observe energy at work. Some of these ways are heat, motion, sound, and light.

**Electric current** comes from an **energy** source such as a D-cell (**battery**) or a wall socket that is connected through wires to a different energy source. When a lightbulb is connected to a source of electricity, the lightbulb will give off light.

When you use a D-cell to light a lightbulb, metal wires carry the electricity from the D-cell to the lightbulb. But if you try to get the lightbulb to light without forming a **complete circuit**, the lightbulb will not shine.

An amusement park is an energy showcase. The thrilling rides, flashing lights, and carnival music are evidence of different forms of energy at work.

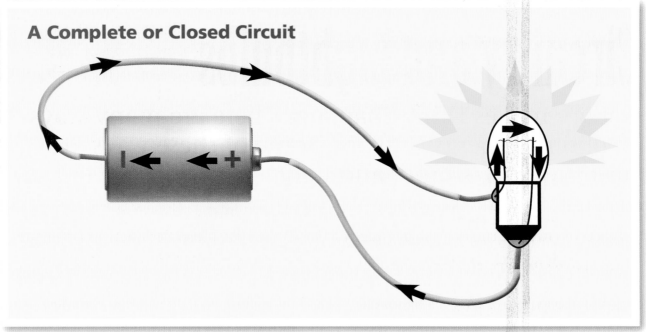

## A Complete or Closed Circuit

This simple circuit has three parts: an energy source (battery), a path for the flowing current (wires), and a device that converts the electricity into a useful form of energy, light.

## Completing the Circuit

The trick is to complete the **circuit**. Electricity needs to flow from the battery *to* the lightbulb, and it also needs to flow *from* the lightbulb back to the battery. One wire connects the base of the lightbulb to one end of the D-cell. The second wire connects the metal casing of the lightbulb to the other end of the D-cell. This setup results in a shining lightbulb. The wires connect the **component** and energy source to form a complete circuit, or a **closed circuit**. The places on a D-cell and lightbulb where wires touch the component are called **contact points**.

If you disconnect one of the wires from the lightbulb or from the D-cell, the electric current pathway is broken, and the lightbulb will stop shining. A circuit with a break is called an **incomplete circuit**, or an **open circuit**. In a standard lightbulb, the lightbulb casing forms one contact point and the base forms another. The two areas are separated by insulating material.

It is important to check which parts of the wires connect to the D-cell and the lightbulb. The wire ends must show exposed metal. Metal conducts electricity, so it is called a **conductor**. Electricity cannot flow through the plastic insulation of the wire.

One exposed metal end of a wire must connect to the positive (+) end of the cell. The metal end of another wire must connect to the negative (–) end of the cell. The other end of one wire must connect to the metal casing of the lightbulb. The other end of the second wire must connect to the base of the lightbulb. These connections make a closed circuit. Only then will electric current flow and the lightbulb will shine.

## Electric Lamps

Lightbulbs are examples of **lamps**. A lamp is simply a source of light. Before the development of electricity, lamps burned oil to produce light. Oil was the energy source for these lamps, not D-cells. A cloth wick was placed in a container of oil and lit on fire. The lamp burned until all the oil was used up.

The same is true for electric lamps, in a way. They burn until the supply of electricity is used up. By running an electric current through the lamp, light shines forth. But how does the lamp produce light?

Remember that for electricity to flow, there must be a complete circuit. Look at the picture and follow the circuit's path. An insulated copper wire enters the bottom of the lightbulb. A metal support wire connects to the copper wire. It clamps onto one end of a thin structure called the **filament**. The other end of the filament is clamped in a second metal support, which connects to the other copper wire that passes out of the lamp. That is the pathway the electricity will take through the lightbulb.

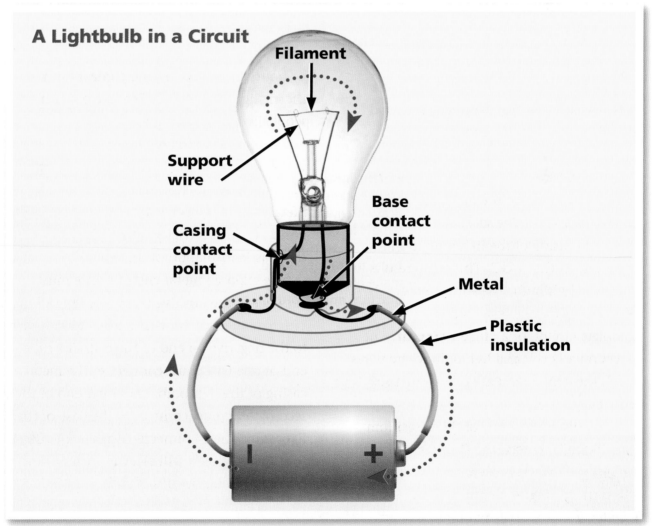

**A Lightbulb in a Circuit**

Filament

Support wire

Casing contact point

Base contact point

Metal

Plastic insulation

**A lightbulb produces light when electric current flowing through the lightbulb's thin filament causes it to get hot enough to glow.**

## Designing Early Lightbulbs

When the lamp is connected to a battery (or other source of electricity), current flows through the circuit, including the filament. The filament is designed to heat up when electricity flows through. It gets so hot that it glows, producing visible light. This is an **incandescent lightbulb**.

It was hard to find a material that could give off light without melting or catching fire. Thomas Edison (1847–1931) and his colleagues made thousands of materials into filaments and ran electricity through them. All the filaments went up in flames after one short burst of light.

Help came from Lewis Latimer (1848–1928), an experienced inventor. He had been working on the engineering problem of the filament. Latimer discovered that a carbon-coated cotton thread made a good filament. He got a patent for the carbon filament. Inventors get patents from the government when they invent something new, so that other people cannot make money off the idea without giving credit to the inventor. When Edison tried the carbon filament in his lab, he agreed that it was the best material. He bought the patent from Latimer. Latimer later joined Edison's laboratory team. He continued to write patents for new inventions.

Another major breakthrough was the idea of sealing the filament in a glass bulb from which the air had been removed. If no oxygen was around the filament, it would not burn up. The idea was successful, and in 1878, Edison produced an incandescent lightbulb with a carbon filament that lasted for hours. He understood the importance of electric lighting. It could change how many people lived.

Electric industry pioneer Louis Latimer was chief draftsman and patent expert for Thomas Edison. His creative breakthroughs helped make electric lighting practical and affordable.

Thomas Edison, widely regarded as America's greatest inventor, was the driving force behind the lightbulb, the phonograph, and the motion picture camera. His Menlo Park, NJ, facility was the world's first research and development lab.

## Lightbulbs Today

In Edison's time, the only way to make electric light was to make a filament so hot that it glowed. The glowing filament gave off a lot of heat and a good amount of light. It takes a lot of energy to make light by heating a filament. Today, we have other ways to make light that do not need nearly as much energy.

The long white tubes used in many classrooms are called fluorescent lamps. Fluorescent lightbulbs were invented in the early 1900s. A fluorescent lamp does not have a filament. Instead, the tube is filled with gas. When an electric current travels through the lightbulb, the gas begins to give off light. The amount of energy needed to produce the light is far less than the energy needed to heat a filament.

In 1976, compact fluorescent lightbulbs were invented. The tube is much thinner, and it is wound into a coil to save space. Compact fluorescent lightbulbs screw into standard sockets designed for incandescent lightbulbs. Replacing all the incandescent lightbulbs in a house with compact fluorescent lightbulbs can save several hundred dollars a year.

Compact fluorescent lightbulbs (CFLs) use about 70 percent less energy than incandescent lightbulbs.

Old-style filaments produce light but also give off a lot of energy as heat.

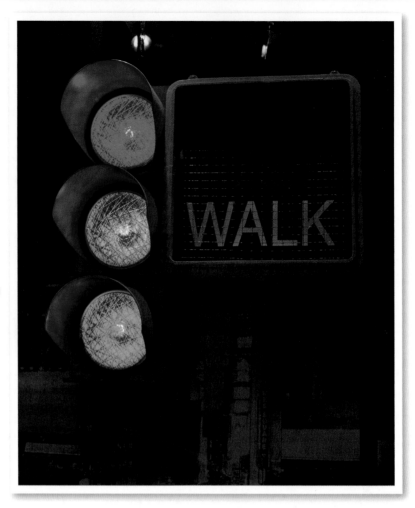

A major advantage of LEDs for traffic lights is their extremely long life—an LED light can last 100,000 hours! In addition, they contain no toxic chemicals, save energy, and are very bright.

In 1962, another light-producing technology was developed. Light-emitting diodes (LEDs) produce light using a small amount of energy and produce very little heat. This makes them extremely energy efficient. But the first LEDs were dim and produced only red light.

Electrical **engineers** developed more powerful and brighter LEDs in a variety of colors. They made bright red, amber, and green lights for traffic lights that save cities a lot of money. Bright white LEDs are now used in streetlights, saving even more energy and money. You may also have seen flashlights that use clusters of small bright lights instead of a single lightbulb. Those small bright lights are LEDs.

For lighting homes and businesses, modern LED lighting uses much less electricity than older technologies. These energy savings are good for the environment, because most electric power plants release pollutants and contribute to **climate change**.

## Think Questions

1. **How do you know when electric current is flowing in a lightbulb circuit?**
2. **Describe the path taken by electricity through an incandescent lightbulb.**
3. **What are some ways in which modern lightbulbs help save energy?**

# What Is Electricity?

Have you ever rubbed a balloon on your hair to hang the balloon on the wall? Have you ever felt "static cling" in your clothes? Have you ever felt a shock after walking across a carpet and then touching something?

These experiences involve a **static** electric charge. Electric charge is a basic property of all matter. Even atoms, which make up everything, have electric charge. Thinking about what is happening at the atomic level helps us understand electricity.

Static electricity is giving this girl a wild hairstyle! Especially in dry winter weather, hairs become positively charged and repel one another, or stand on end.

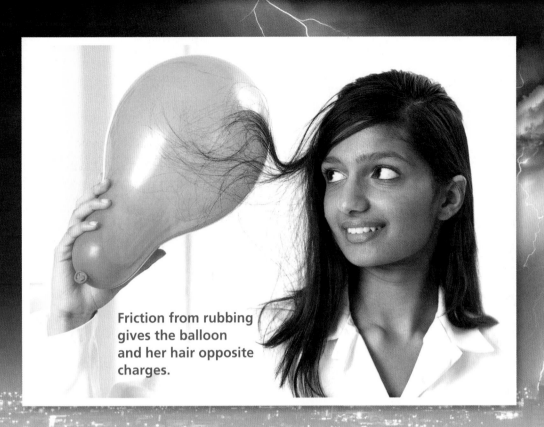

Friction from rubbing gives the balloon and her hair opposite charges.

## Electric Charge

**Electrons** form the outer part of the atom. Each electron has a negative electric charge. If an electron breaks away from an atom, that atom is left with a positive electric charge.

When you rub a balloon on your hair, some electrons break away from the balloon's

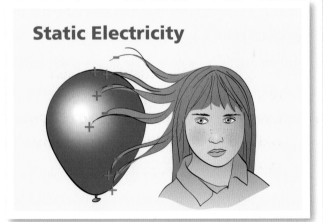

### Static Electricity

When electrons move from the balloon to the hair, the balloon is left with a positive charge and the hair gains a negative charge. Now the balloon and the hair attract each other!

atoms. Now the balloon has a slightly positive electric charge. Where do the electrons go? They join the atoms of your hair. They give your hair a slightly negative electric charge.

Just as magnetic poles attract or repel, electrically charged objects can attract or repel.

- Opposite charges attract. Things with opposite charges, like your hair and the balloon, attract each other.
- Like charges repel. Things with negative charges repel each other. Things with positive charges repel each other.

When your hair sticks up because of static electricity, the hairs are positively charged. Each hair is being repelled by the other hairs!

The force between two charged objects is an **electric force**, another of the invisible forces of nature like gravity and magnetism.

Lightning is a spectacular discharge of electricity in the atmosphere, usually during thunderstorms. Scientists estimate that 44 lightning bolts strike Earth's surface every second.

## Electric Current

Electric charge can be static, which means not moving. Electric charge can move, which is called electric current. Remember getting a shock after walking on a carpet and touching an object? As you walk, static electric charge builds up on your body. If you have enough static electric charge, a spark will jump from your finger to the object. The spark is actually moving electric charges.

Remember your work with the lightbulb, battery, and wire? Electric charges will not move unless there is a complete circuit. Think of electric force as a push. The force pushes electrons through the circuit's pathway. If the circuit is not complete, the electric current will not flow.

Lightning is also a short-lived electric current, but on a huge scale. The bottoms of clouds build up negative charges, and the ground builds up positive charges. The lightning bolt is a giant spark as the electric current moves to balance the charges. This powerful spark is very dangerous for anyone in its path.

### Did You Know?

It's best to seek shelter at the start of a storm, before lightning begins. Electricity travels at the speed of light, which is nearly 300 million meters per second!

## Using Electricity

We use electricity every day to power lightbulbs, TVs, computers, light-rail trains, and electric cars. This electricity consists of electrons moving through wires. The wires are typically made of the metal copper. They are coated with insulating material such as plastic.

We use wires to connect an electrical energy source to a device. The source could be a battery or an electric outlet. The electric force is instantly felt throughout the wire, pushing electrons along the path. Electrons flow through the circuit, creating an electric current.

Why is copper used to make electric wires? The arrangement of electrons in copper atoms makes it easy for them to produce an electric current. Other metals, like silver, gold, steel, and tin, have the same property. They also conduct electricity well. Nonmetals have different electron arrangements and may conduct electricity poorly (called **semiconductors**). Some do not conduct at all (**insulators**). The plastic coating on a copper wire is a good insulator.

Out of sight behind walls, ceilings, and floors is a network of wires and cables that make your world run smoothly. To carry current safely, the metal conductors are wrapped inside plastic insulators.

## Battery Structure

A battery is an energy source. Energy cannot be created or destroyed. But it can transfer from one form to another or from one place to another. Batteries change chemical energy into electrical energy.

The most common disposable batteries have these essential parts.

- The positive (+) terminal touches a graphite rod surrounded by a chemical mixture (electrolyte).
- The negative terminal (–) is a zinc container.
- The electrolyte transfers electric charge between the positive and negative terminals.
- A thin layer of paper or fabric separates the electrolyte from the zinc container.

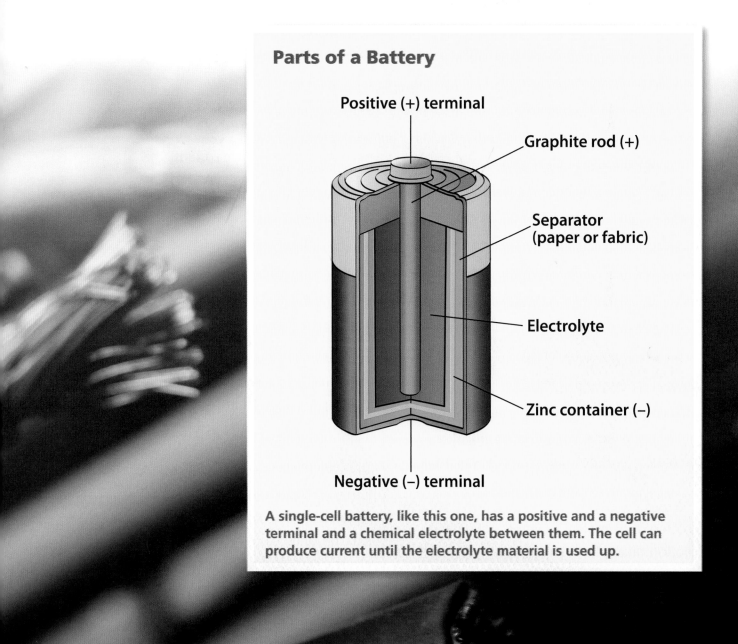

### Parts of a Battery

Positive (+) terminal

Graphite rod (+)

Separator (paper or fabric)

Electrolyte

Zinc container (–)

Negative (–) terminal

A single-cell battery, like this one, has a positive and a negative terminal and a chemical electrolyte between them. The cell can produce current until the electrolyte material is used up.

The electrolyte inside most single-use disposable batteries is alkaline, which means it has a high pH. If a battery breaks open or leaks, the chemicals can cause severe damage to skin and eyes.

## How a Battery Works

A typical D-cell has two terminals, positive (+) and negative (–). These terminals must be connected to complete a circuit with the battery. The chemical reaction inside the battery will not start until the circuit is complete.

When a wire connects two terminals of a battery to complete a circuit, a chemical reaction takes place. The reaction releases electrons from the zinc atoms. Electrons have a negative charge, so they flow through the circuit toward the positive terminal. Electric current passes through the entire circuit, including components like lightbulbs, which light up as the current passes through.

As the reaction progresses, the chemicals get used up. Eventually, the chemical reaction stops. The electric current stops flowing. The battery is dead and ready for proper disposal.

Rechargeable batteries are used in smartphones, cameras, laptop computers, electric cars, and many other devices. The most common rechargeable batteries today are lithium ion batteries.

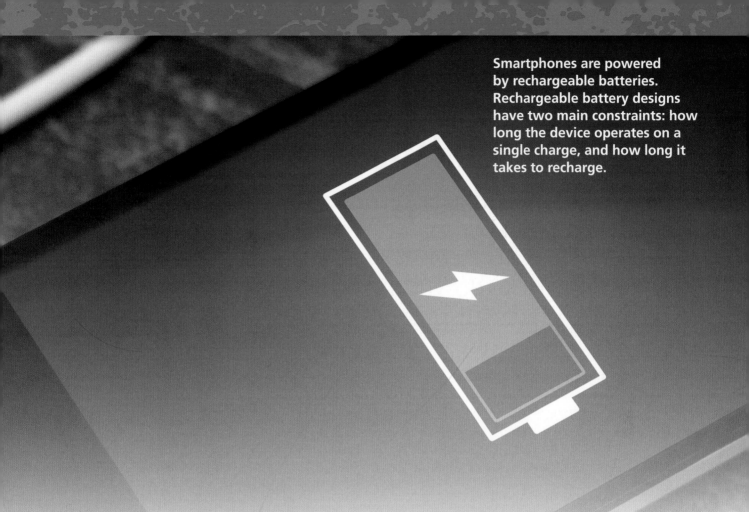

Smartphones are powered by rechargeable batteries. Rechargeable battery designs have two main constraints: how long the device operates on a single charge, and how long it takes to recharge.

Rechargeable batteries work like a disposable battery. However, connecting a rechargeable battery to an electrical energy source reverses the chemical reaction. When the battery is charged, the reaction can take place again. Rechargeable batteries eventually wear out, but only after hundreds of recharging cycles.

## Did You Know?

All batteries contain chemicals that can harm the environment. A discarded battery breaks down over time and can pollute the local ecosystem and ground water. Many stores and community organizations accept used batteries for safe disposal or recycling. Find out where to recycle batteries in your community.

## Think Questions

1. If you rubbed two balloons on your hair to charge them electrically, would the two balloons attract or repel one another?
2. If you rub a balloon on your hair, you can hang the balloon on the wall. Why does the balloon stick to the wall?
3. Do all materials conduct electricity equally well?
4. Why are insulators important?

# Electromagnetism

A massive junk heap sits, a mixture of trash and valuable materials. A giant magnet pulls steel scraps out for recycling. How do you unstick the steel from a powerful magnet? The solution is a temporary magnet that you can turn on and off.

## The Discovery of Electromagnetism

The battery was the first source of electric current. It was invented by Alessandro Volta (1745–1827) in 1800. The D-cell we use today is a direct result of Volta's discovery.

The magnetic property of steel and iron makes these metals the easiest materials to separate from the solid waste stream. Millions of tons of steel are diverted from trash to reuse each year, pulled from scrap heaps by powerful electromagnetic cranes.

Hans Christian Ørsted (1777–1851) was a Danish physics professor in the late 1700s. He was fascinated by Volta's battery. Ørsted conducted many experiments with electric current.

In 1820, Ørsted was demonstrating that electric current makes wires hot. When he closed the electric circuit, the needle of a compass on the lecture table rotated. Some people think that Ørsted had planned to show the relationship between electric current and magnetism that day. Others think it was just a lucky accident. We will never know for sure.

Here is what might have happened. Ørsted had a thin wire connected to a battery and a switch. A compass needle was right under one of the wires forming the circuit.

When Ørsted closed the circuit to deliver electric current to the thin wire, the needle rotated.

When Ørsted made this discovery, he conducted more experiments. Four months later, he wrote about his findings. He concluded that a flow of electric current produces a magnetic field.

This important discovery resulted in hundreds of inventions in the years that followed. One was the **electromagnet**, a magnet that can be turned on and off.

## The Discovery of Electromagnetism

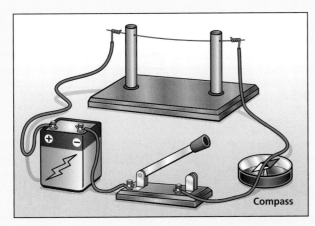
Compass

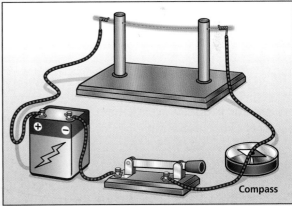
Compass

Whether by accident or by design, Ørsted discovered the relationship between electricity and magnetism. Both are caused by the electromagnetic force.

## Electromagnetism

The wire you have been using to make circuits is made of copper. Copper is not magnetic. There is no magnetic field around a copper wire. You can confirm or prove this by bringing a compass close to a copper wire. The compass needle does not move.

Things change when you connect a copper wire to a source of electricity, such as a D-cell. While electric current flows through the wire, a magnetic field surrounds it. When you bring a compass close to the wire, the compass needle rotates slightly. When you break the circuit, the magnetic field disappears. The compass needle points north again.

The magnetic field around a wire that has electric current flowing through it may not be very strong. But if you put two magnetic fields together, the magnetism becomes stronger. That is what happens when you coil up a wire. The magnetic field around each loop of wire adds to the fields from other loops. The greater the number of loops, the stronger the total magnetic field is.

### Electric Current and Magnetic Field

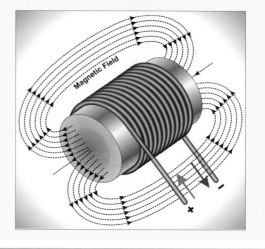

Magnetic Field

Copper itself is not magnetic. But the flow of electric current through a copper wire creates a magnetic field around the wire.

## Magnetic Fields of Electromagnets

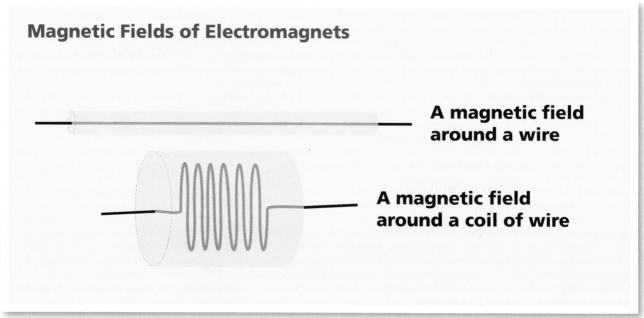

**A magnetic field around a wire**

**A magnetic field around a coil of wire**

Increasing the number of coils or loops of wire increases the strength of the magnetic field.

Steel is the world's most recycled material, more than paper, plastic, glass, and aluminum combined. Yesterday's cars and toasters may be tomorrow's bridges and skyscrapers.

What happens when the coil wraps around a steel **core**, like a rivet? The strong magnetic field induces magnetism in the steel. The steel becomes a temporary magnet as long as the electric circuit is complete. And when you open the switch, the magnetism turns off. This is **electromagnetism**.

Not long after Ørsted's discovery, Michael Faraday (1791–1867) discovered that magnetism could be used to create electric current. From then on, it was clear that one force was behind both magnetism and electricity. That force is the **electromagnetic force**.

Now you know how metal objects can quickly be sorted for recycling. Strong electromagnets are used in recycling centers for separating some metals (mostly iron and steel) from other scrap metal. Aluminum cans, for example, are left behind because they are not attracted to magnets. Turn on a large electromagnet over a junk heap, and watch as it lifts steel parts from the pile. Move the electromagnet over your collection bin. Then break the circuit. When the current stops flowing, the electromagnet no longer has induced magnetism. The steel falls into the bin below.

## Think Questions

1. **What was Ørsted's historic discovery?**
2. **How do you make an electromagnet?**
3. **How do you think you could make an electromagnet stronger?**

# Electromagnetic Engineering

## This train is off the tracks. As it moves, it hovers just above the tracks.

A train can travel more than 400 kilometers (km) per hour. It is propelled by electromagnetic force. How did the first electromagnets develop into an electromagnetic train?

### Communicating with Electromagnets

Countless scientists and engineers have used electromagnets in their designs. Some designs use simple electromagnets combined with other components to transmit messages.

In 1835, Samuel F. B. Morse (1791–1872) used an electromagnet, a switch, and a battery to send a long-distance message. Long wires ran to an electromagnet far away. When the sender pressed a switch to complete the circuit, the receiver's electromagnet attracted a piece of steel and made a loud click. Those

The Shanghai maglev train was the world's first commercially operated magnetic levitation line. Now several countries (but not the United States) operate these high-speed, reduced-friction "bullet" trains.

first clicks announced the invention of the telegraph.

Electromagnets are essential in the telephone, too. Alexander Graham Bell (1847–1922) invented the telephone in 1876. To hear sounds over a phone, you need a speaker to convert electric signals into sound vibrations. Most speakers have a coil of wire attached to a paper cone. The coil of wire is near a permanent magnet. When electric signals pass through the coil, the electromagnetic field of the coil interacts with the permanent-magnet field. This makes the paper cone vibrate, creating sound waves in the air. Headphones and earbuds work much the same way, but use much smaller magnets and coils of wire.

The telegraph key is a switch that opens and closes a long-distance electric circuit, transmitting signals—the dots and dashes of a coded alphabet—by wire.

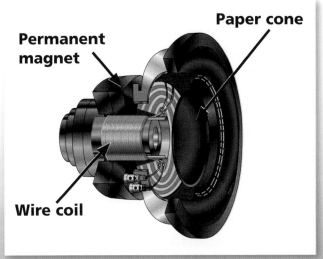

Speakers come in all shapes and sizes, from compact earbuds to stadium loudspeakers. All use electromagnets to change electric energy into sound energy.

**Permanent magnet**

**Paper cone**

**Wire coil**

Microphones convert sound into electric signals. Some music-recording microphones use electromagnets. For example, a dynamic microphone acts like a speaker in reverse. Instead of a paper cone, a dynamic microphone has a disk. A coil of wire attached to the disk sits in the field of a permanent magnet. When the sound waves make the disk vibrate, the coil vibrates in the magnetic field, generating electric vibrations.

## Other Designs That Use Electromagnets

Model railroads sometimes use electromagnets for switching tracks. Tiny electric **motors** physically move the track. Electric motors are related to electromagnets, as you will see later in this course.

Induction heating uses electromagnets. An induction stovetop stays cool on its surface and heats only pans made of iron or steel. When the iron pan is on the stove, the stove converts electric current into thermal energy to heat the pan. Induction heating is also used in manufacturing and medical treatments.

An induction stovetop is safe to touch while cooking. The burner heats your pot but not your palm.

**Take Note**

**Why do you think the induction stove heats only pans containing iron?**

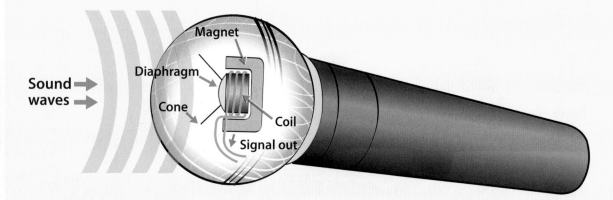

## Dynamic Microphone

Sound waves

Magnet

Diaphragm

Cone

Coil

Signal out

Dynamic microphones use electromagnets to change sound to electricity. The current can then be sent to an amplifier to be changed back to (louder) sound. Or it can flow to a recording device to be stored.

## Modifications to Electromagnets

How can you make an electromagnet stronger? One way, which you may have figured out, is to have more coils of wire. The more wire you wrap around the core, the stronger the magnetism.

Another way to make an electromagnet stronger is to increase the amount of electric current flowing through the wire. You used one D-cell for your electromagnet.

Two D-cells in a series circuit increases the magnetism a lot. What if you had ten D-cells in series? Or a giant car battery? Now we're talking about some strong magnetism!

A third way to make an electromagnet stronger is to wrap thicker wire around the core. Thicker wire can conduct more current. The thicker the wire is, the stronger the magnetic field, and the stronger the electromagnet.

**Electromagnets have two important features that permanent magnets lack: Their magnetism can be turned on and off, and their magnetic fields can be strengthened with more coils and thicker wires.**

## Superconducting Magnets

Some of the strongest electromagnets, called superconducting magnets, are cooled to reach maximum strength. Their coils of wire are cooled to extremely low temperatures, about –269 degrees Celsius (°C). Supercooled wires can conduct much larger electric currents than ordinary wire. This allows them to create intense magnetic fields. These magnets are efficient to operate, because no energy is lost heating the wire. Magnetic resonance imaging (MRI) machines in hospitals use such magnets.

Superconducting magnets can be strong enough to lift a whole train above its tracks. The Shanghai **maglev** train is a high-speed magnetic-levitation train. It carries passengers from the airport in Shanghai, China, to the city center. The 30 km (18.6 mi) trip takes less than 8 minutes.

The term *maglev* combines the words *magnetic* and *levitation*. Electromagnetic force levitates the cars between 8 and 12 millimeters (mm) above the track. The levitating cars do not have friction with the track. The cars still have friction with air, however. This air resistance is called **drag**. An aerodynamic design minimizes drag. Once the train levitates, superconducting electromagnets alternately push and pull the train along the track, faster and faster. The Shanghai maglev train reaches speeds over 500 km per hour (300 mph). Without friction on the tracks, the ride is extremely smooth.

### Think Questions

1. How has use of electromagnetic technology changed over the years?
2. Pick out three devices described in this article. List an engineering constraint or criterion that would be important to consider in designing such a device.
3. What did you learn from this article that could make your electromagnetic design stronger?

MRI scanners use extremely powerful superconducting magnets, radio waves, and computers to produce images of tissues inside a person's body. The images help doctors diagnose conditions like brain tumors.

Charging stations are evidence that electric car sales are trending upward. Cautious consumers may choose a hybrid, with both a gasoline engine and an electric motor for different driving needs.

# The Rebirth of Electric Cars

Walking through the parking lot at a mall or movie theater, you might see parking spaces reserved for electric cars.

These special parking lots are becoming more common as more drivers use electric cars. Drivers can recharge their car while they go shopping, have lunch, or watch a movie. Electric cars were invented in the 1800s, so why are they becoming popular now?

**Did You Know?**

The first electric cars were model cars built in the 1820s and 1830s. It was not until the 1880s that fully functional full-scale electric cars were built.

## The First Automobiles

Before the early 1800s, road vehicles were powered by humans or animals, often horses. Then, engineers began to design motors that would move vehicles. They named these machines **automobiles** (self-movers).

An automobile needs an energy source. Early automobiles burned coal or wood. They heated water until it boiled. The resulting steam powered the automobile's engine.

A wood stove and water boiler were heavy and took up a lot of space. So engineers developed automobiles that used different energy sources. Vehicles powered by gasoline or diesel (another petroleum product) rely on a series of little explosions inside the engine, which is why they are called internal combustion engines. Electric automobiles are powered by batteries.

In the early 1900s, steam-powered and electric automobiles were the most popular. By 1915, petroleum was widely available. Gasoline-powered automobiles became popular. They ruled the roads throughout the rest of the 20th century and the beginning of the 21st century.

Early electric cars, like this 1910 model, were quiet, clean, easy to drive, and favored by urban drivers. As gasoline became widely available, however, they were replaced by larger, more powerful gasoline-powered vehicles.

## Redesigning Electric Cars

It took decades of engineering to develop electric cars that could compete with gasoline-powered cars. To understand the design challenge, consider the three main components of the electric car: the electric motor, controller, and battery. When you switch on the car, the battery transfers chemical energy into electrical energy. The controller transfers electricity from the battery to the motor. The motor then transfers electrical energy to mechanical energy.

## A Brief History

**1837**
Scottish chemist Robert Davidson (1804–1894) builds the first electric locomotive.

**1859**
French physicist Gaston Planté (1834–1889) invents the first rechargeable battery for storing electricity on a vehicle.

**1899**
An electric racing car sets the land-speed record at 109 kilometers (km) per hour.

**1900**
About 38 percent of all registered automobiles in the United States are electric (33,842 automobiles). About 40 percent are steam-powered, and 22 percent are gasoline-powered.

**1915**
Only 5 percent of automobiles are electric in the United States. Ford Motor Company's™ successful Model T gasoline-powered car has changed the industry.

**1942**
The military operations of World War II are possible because of petroleum-powered cars, trucks, tanks, ships, and planes.

**1971**
The first manned vehicle on the Moon is an electric lunar rover.

**1973**
Gasoline prices rise, as does interest in electric cars. The US Department of Energy funds efforts to make a cost-effective electric car.

**1996**
General Motors™ releases the EV1™. Production costs are high, and General Motors™ abandons the project in 2001.

**2008**
The Tesla Roadster™ is the first successful all-electric car. It has a range of over 200 km.

**Today**
Advances in technology make electric cars more and more practical.

## Defining the Problem: Criteria

Engineers use their scientific knowledge to design solutions to real-life problems. They ask, "What are the **criteria** that define a solution to the problem?" The greatest challenge for a successful electric car is the relationship between two criteria: vehicle performance and vehicle range.

### Criterion: Performance

Creating high-powered electric motors has been possible since the early 1900s. In fact, electric vehicles held the land-speed record until the early 1900s. The 2012 Tesla Roadster™ can accelerate from 0–96 km per hour in under 5 seconds. That power requires a lot of energy. The more energy you want to use at once, the faster the battery drains.

### Criterion: Range

The range of a car is the distance it can travel on one battery charge. High speeds use lots of energy, which reduces the range of the car. Heavier cars require more energy to move, which also reduces the range. This is why most electric cars are small vehicles.

## Defining the Problem: Constraints

Another important question asked by engineers is "What are the limits, or **constraints**, on a potential solution?" You may have already guessed one constraint, based on the information about speed and range.

Look under the hood of an electric car. Instead of pistons and cylinders, pumps and belts, you'll find a hefty battery pack. The design criteria are to create a safe, affordable battery whose size and weight give a vehicle both performance and range.

## Constraint: Battery

A battery is a portable source of stored chemical energy. Batteries produce electric current when their chemicals react. When the chemicals are used up, the battery no longer produces electric current.

Electric cars need batteries that can be recharged. Some batteries can be recharged by running an electric current through them. This reverses the chemical reaction, changing the chemicals back to their starting condition. The recharged battery is again ready to produce electricity. As engineers work on new designs, they try to increase the battery range while decreasing battery size and weight. Three kinds of batteries have been used in electric cars. Each type contains chemicals that store and release energy through a reaction.

Lead acid batteries were first invented in 1859. These are the oldest rechargeable batteries. They are cheap, used in many machines, and 97 percent recyclable. The main constraint of lead acid batteries in electric cars is that they are very heavy for the amount of electric power they produce. This ratio limits the range of the car.

Nickel metal hydride batteries are more expensive than lead acid batteries. But for the same size and weight, they can provide more energy.

Lithium ion batteries were first used for portable electronics like laptops and mobile phones. Lithium ion batteries led to major design changes in electric cars in 2009. Lithium is the third lightest of all the elements, far lighter than nickel or lead. The chemicals in these batteries are more expensive to manufacture than those in other batteries.

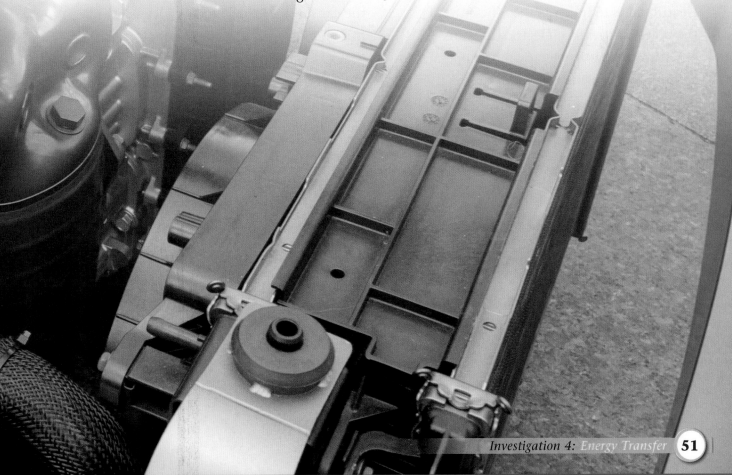

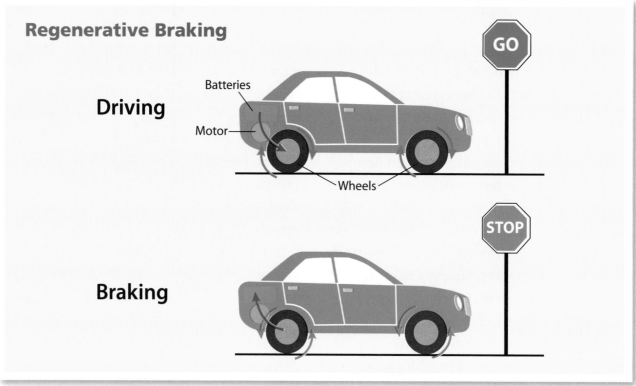

**Regenerative Braking**

Driving

Batteries

Motor

Wheels

GO

Braking

STOP

In regenerative braking, the motor acts as a generator, producing electricity as the car slows down and stops. This energy is fed back into the onboard storage batteries to recharge them.

## Constraint: Range

Another way to extend the range of an electric car is to increase the efficiency of the power system. An electric car gains efficiency when the electric motor operates as a **generator** during braking.

What happens when a driver brakes? The force of friction on the brake pads converts motion energy to thermal energy. Energy is lost to the environment through heating as the car slows. During regenerative braking, this energy is not all lost. The motor turns in the opposite direction to become a generator. It converts the motion energy of the car into electrical energy. That energy recharges the batteries while acting as a brake for the car.

### Electric Automobile Designs and Performance

| Year | Make | Model | Battery type | Top speed (km per hour) | Range (km) | Base price[1] |
|---|---|---|---|---|---|---|
| 1907–1939 | Anderson Carriage™ | Detroit Electric™ | Lead acid | 32 | 130 | $2,600 (~$64,000) |
| 1996 | General Motors™ | EV1™ | Nickel metal hydride | 129 | 257 | $34,000 (~$48,600) |
| 2008 | Tesla™ | Roadster™ | Lithium ion | 201 | 322 | $109,000 (~$120,500) |
| 2011 | Nissan™ | Leaf™ | Lithium ion | 150 | 160 | $35,400 (~$37,600) |

[1]Price in US dollars at time of sale (price in 2016 dollars).

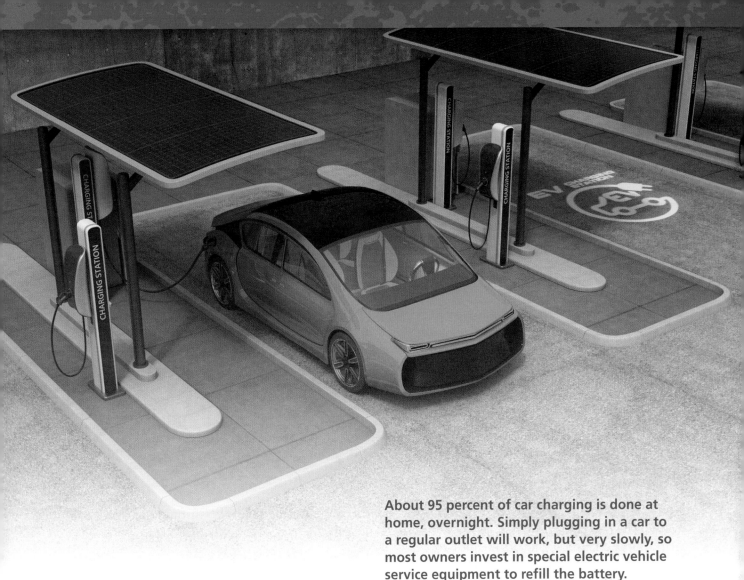

About 95 percent of car charging is done at home, overnight. Simply plugging in a car to a regular outlet will work, but very slowly, so most owners invest in special electric vehicle service equipment to refill the battery.

## Benefits of Electric Cars

Electric car use is on the rise. The main drawback is the limited range. For many drivers, however, a range of 110–130 km is enough for normal daily use. For occasional longer trips, drivers carefully plan stops at charging stations or use a gasoline car instead.

People who drive electric cars appreciate these features.

- They never need a gas station.
- The electric car has low maintenance costs. The power system has fewer parts than a gasoline-powered car.
- Charging can use **sustainable** energy sources.

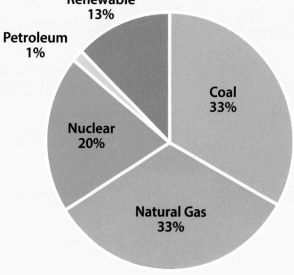

## Energy Sources for US Electricity in 2015

Renewable
13%

Petroleum
1%

Coal
33%

Nuclear
20%

Natural Gas
33%

About two-thirds of the electricity we use is generated by burning fossil fuels. Even clean, "green" electric cars rely mostly on this energy resource, although far less than cars powered by combustion engines.

## Sustainable Energy

Sustainable means something that can be maintained at a certain level, or **renewable**. Gasoline is not sustainable, because when it is used, it cannot be replaced. Gasoline and other **fossil fuels**, like coal, natural gas, and diesel, come from decomposed plant material that was buried. The material changes chemically underground over hundreds of millions of years. It becomes substances that can be burned as **fuel**. Because fossil fuels take a long time to form, we use them up more rapidly than they can form. They are not a sustainable energy source.

When cars burn fossil fuels, they release pollutants into the atmosphere. Air pollution is hard on human health. Pollutants can also affect the conditions of the atmosphere. More and more scientific evidence indicates that carbon dioxide from burning fossil fuels is

causing climate change. Climate change leads to serious problems around the globe.

Because of the fossil-fuel problems, some consumers are turning to electric cars. The cars emit no exhaust, are more energy efficient (require less fossil fuel), and could be recharged with sustainable sources of energy like solar and wind power.

Right now, most electric cars are charged with electricity that was generated at a power plant. As represented in the chart above, in 2015, 33 percent of US electricity was generated by burning coal. About 33 percent was generated by burning natural gas. About 20 percent came from nuclear power. About 1 percent came from burning petroleum, which is used to make gasoline and oil. The remaining 13 percent came from renewable sources like hydroelectricity and solar and wind power.

## The Future of Electric Cars

Challenges for electric-car users are the lack of charging stations, the length of time it takes to recharge a car, and short driving range. Engineers are trying to create faster charging stations that generate more electricity from sustainable energy sources. Some engineers are looking into the idea of battery-changing stations. The driver would pull in. The car's whole battery pack would be removed for recharging and replaced by an already-charged battery pack.

As more people want electric cars, more manufacturers are designing and producing the cars. This in turn helps lower the cars' cost.

### Think Questions

1. **Why were gasoline-powered automobiles so popular for most of the 20th century?**
2. **What are some technological limitations for electric cars?**
3. **How do electric cars help reduce fossil-fuel use?**

One hurdle facing engineers is recharging time: several hours for a full recharge. Public "quick chargers" that can add about 80 km of range in a half-hour are becoming increasingly available.

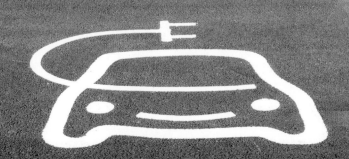

Without electrical energy, just about everything in our day-to-day lives shuts down. Our reliance on electricity is a part of modern life.

# Where We Get Energy

The lights gleam from the city at night. You can see giant buildings lit head to toe, and watch lines of cars driving along. People walking around feel safe in the brightly-lit streets. Humans rely on energy to light their towns and drive their cars.

One definition of energy is "the ability to apply force over a distance." That does not really capture how much we rely on energy. Energy makes things happen. It supports nearly everything we do. We use energy to move cars and buses, heat our homes, cook food, and power computers. A big challenge for our society is how to get sustainable energy to power our daily life.

## The Greenhouse Effect

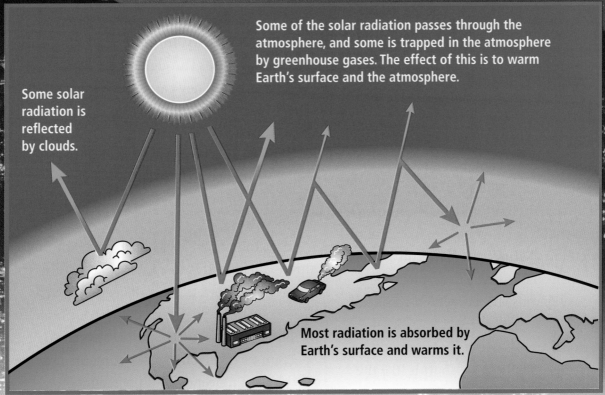

Some solar radiation is reflected by clouds.

Some of the solar radiation passes through the atmosphere, and some is trapped in the atmosphere by greenhouse gases. The effect of this is to warm Earth's surface and the atmosphere.

Most radiation is absorbed by Earth's surface and warms it.

Gases in the atmosphere keep our planet warm, making life possible. But the extra greenhouse gases added by global human activity are changing Earth's climate.

## The Fossil-Fuel Source

Fuel is a material that stores energy. Our ancestors were burning wood as fuel more than 400,000 years ago. Humans have been burning coal as fuel for over 5,000 years. In the past 100 years or so, humans have grown used to cheap energy from fossil fuels. Fuels such as coal, oil, and natural gas have been plentiful and inexpensive. Using fossil fuels, however, creates serious problems.

When we burn fossil fuels to release energy, a chemical reaction occurs. The reaction uses the stored chemical energy of the fuel to generate thermal energy. Thermal energy can be transferred to mechanical energy and electrical energy. Mechanical energy powers the engines of cars, buses, and planes. Electrical energy travels from a power plant to our communities through the **power**

**grid**. A power grid is a complex network of wires that delivers electricity to homes and other buildings.

Burning fossil fuels releases carbon dioxide and other pollutants into Earth's atmosphere. Carbon dioxide is a **greenhouse gas**. It helps the atmosphere "trap" thermal energy that would otherwise radiate into space.

As carbon dioxide builds up in the atmosphere, Earth becomes warmer. This is known as climate change. Human-induced climate change is driving global temperatures higher and altering climate all over Earth. Climate change leads to serious problems around the globe, like sea level rising, more severe droughts, more severe storms, and heat waves. Carbon dioxide emissions are also causing the ocean to become more acidic, endangering ocean life.

## Powering a Grid

Most of our electricity comes from power plants that burn fossil fuels. How does burning coal, oil, and natural gas make electricity?

Let's look at a power plant that burns coal to heat water. It creates high-temperature, high-pressure steam. That steam rushes through a **turbine** and spins it. A turbine is a machine that has many blades, like a highly engineered fan. Natural gas plants also use turbines. Instead of using steam, burning natural gas expands the gas, which spins the turbine.

The turbine turns a shaft that connects to an electric generator. The shaft is like the hand crank we used in class, only much, much larger. As the shaft rotates, coils of wire interact with permanent magnets to generate electricity. In some generators, this electricity passes through a **commutator** and **brushes**. They carry the electricity from the rotating coils to wires that connect to the power grid.

The power plant transforms the electricity for your home into a high voltage. It can travel a long way with very little energy loss in the wires.

That high-voltage electricity goes through high-voltage wires to a local transformer. The transformer lowers the voltage to 240 volts or 120 volts before delivering it to your home.

Burning fossil fuels produces about 67 percent of US electricity. Coal plants provide 33 percent, natural gas 33 percent, and petroleum about 1 percent.

Turbines are like giant pinwheels. They spin when steam, expanding gas, falling water, or wind turns their blades. A spinning turbine provides the kinetic energy for an electricity generator.

## Alternative Energy Sources

Other sources of heat are used for generating electricity. About 20 percent of US electricity comes from nuclear power. In nuclear power plants, the radioactive decay of the element uranium heats water to make steam and turn turbines. Uranium is mined and is **nonrenewable**, like fossil fuels.

Nuclear power plants do not produce greenhouse gases that would contribute to global warming. But they do produce radioactive waste. That waste must be carefully contained and stored for thousands of years. These plants present a risk to nearby communities if their equipment breaks or is damaged by a natural disaster, like an earthquake.

Fossil fuels and nuclear fuel are not sustainable energy sources. Fossil fuels form over hundreds of millions of years and cannot be replaced as fast as we are using them up. The situation with nuclear fuel is similar.

Burning wood to make steam for turbines is also possible. About 0.4 percent of US electricity is produced that way. Is wood a sustainable energy source? That depends on whether trees are replanted and how fast they can grow.

Nuclear power plants generate electricity by using steam to turn turbines. The heat to make the steam is produced not by burning fuel but by splitting the nucleus of a uranium atom, a process called fission.

Hydropower, produced by the force of flowing or falling water, is the most widely used renewable energy source. Hydroelectric dams are nonpolluting, but they do have significant environmental impacts.

Iceland has lots of geothermal activity. About 25 percent of Iceland's electricity comes from geothermal energy.

## Sustainable Energy Sources

One source of sustainable energy is Earth's heat. In some places, Earth's crust is quite thin. Water just beneath the surface can be very hot. Sometimes, geological features called geysers shoot steam and water out of the ground, or hot springs form. Geothermal power plants in these locations use high-pressure steam or very hot water from deep underground to run turbines. Geothermal power plants do not produce greenhouse gases. They produce only about 0.5 percent of US electricity.

A turbine can spin without needing to burn fuel or collect heated steam. Water flowing through a dam can turn turbines.

The turbine connects to a generator to make electricity. This sustainable method, called hydroelectric power, produces no greenhouse gases. It produces about 6 percent of US electricity.

Hydroelectric power does create problems. Damming rivers and converting them into lakes has serious effects on surrounding ecosystems. As the lake behind the dam fills up with sediment, the water pressure drops. Lower water pressure reduces the efficiency of the power plant. Engineers are working to develop water turbines to harness power from rivers, ocean waves, and tides. These turbines are more efficient and do less harm to ecosystems than turbines in dams.

## Variable Energy Sources

Wind turbines can be connected to generators. This technology is simple, inexpensive, and sustainable. It produces about 5 percent of US electricity. One drawback of wind-electric power is that it varies, depending on wind speed. And when wind speeds are too high, wind turbines must stop to avoid damage. For this reason, wind-generated electricity is usually stored. It is just one source of electricity connected to a power grid.

Photovoltaic technology does not depend on electric generators at all. It uses solar panels made of flat arrays of **solar cells**. When sunlight shines on a solar cell, part of the energy in the light is converted directly into electricity.

You can find solar panels on the roofs of some schools, businesses, parking lots, and homes. Some very large power plants in deserts use solar cells to generate electricity for cities. Only about 0.7 percent of US electricity comes from solar panels. Solar installations are rapidly growing, so the percentage of US electricity that is solar could increase dramatically in years to come.

Like wind power, solar power is variable. It does not generate electricity at night or on dark stormy days. This energy source is sustainable in the long run, but must be stored or added to other sources of electricity. One relatively simple way to store electricity is in rechargeable batteries. Another is using the electricity to pump water uphill. Holding the water in that position stores **potential energy**. To produce electricity, the water flows through a turbine in a hydroelectric system when the stored energy is needed as electricity again.

### Take Note

**Ask your local electric company what energy sources and technologies provide your community's electricity.**

# The Future of Renewable Energy

As we start to experience the effects of climate change, countries around the world are working together to reduce carbon dioxide emissions. Renewable energy sources supply just 13 percent of US power, but it's a start. What are other countries doing?

- Sweden is becoming 100 percent fossil-fuel free. It is investing heavily in solar and wind power and energy storage.
- Germany has turned its attention to solar energy. Depending on weather conditions and time of year, it produces up to 78 percent of its electricity from renewable sources.
- China is committed to phasing out coal burning. It is also addressing its serious air pollution problems. It is the largest producer of renewable energy in the world, mostly from wind and solar power.

The United Nations brings countries together to use more sustainable energy sources. Each year, countries agree to goals that each country works to meet. These goals call on engineers to develop new technologies. They encourage all people to reduce energy use. We can all help reduce carbon emissions and try to slow climate change.

Solar energy is the cleanest and most abundant energy source available. These rooftop panels convert sunlight to electricity that can be stored and used to meet household power needs.

## Think Questions

1. Why is using fossil fuels considered unsustainable?
2. What are some advantages of using sustainable energy sources?
3. Compare the different energy sources described in this article. In your notebook, draw a table like the one below. Mark with an X to show how each energy source works. The first row shows coal as an example.

| Energy source | Uses steam | Uses turbine | Uses generator | Emits carbon dioxide | Sustainable in long term |
|---|---|---|---|---|---|
| Coal | X | X | X | X | |
| | | | | | |

# Images and Data

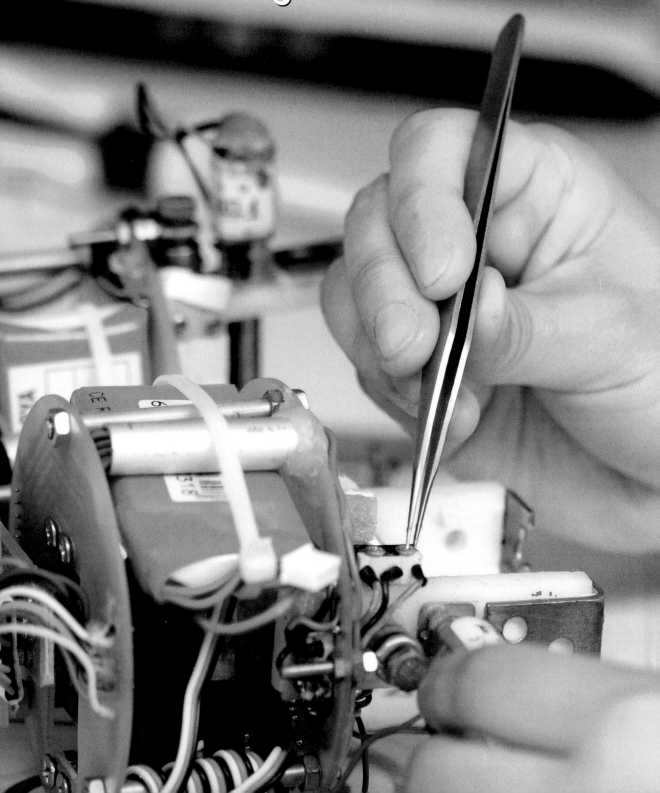

# Images and Data  Table of Contents

**Investigation 3: Electromagnetism**
Parts of an Incandescent Lightbulb . . . . . . **67**
Engineering Design Process . . . . . . . . . . . **68**

**Investigation 4: Energy Transfer**
Motor Dissection A . . . . . . . . . . . . . . . . . **69**
Motor Dissection B . . . . . . . . . . . . . . . . . **70**
Generator Dissection . . . . . . . . . . . . . . . **72**

**References**
Science Practices . . . . . . . . . . . . . . . . . . . . . **73**
Engineering Practices . . . . . . . . . . . . . . . **74**
Science Safety Rules . . . . . . . . . . . . . . . . **75**
Glossary . . . . . . . . . . . . . . . . . . . . . . . . . . **76**
Index . . . . . . . . . . . . . . . . . . . . . . . . . . . . **78**

# Parts of an Incandescent Lightbulb

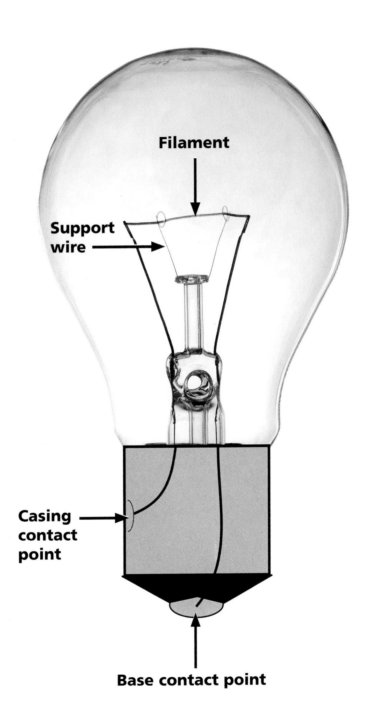

Filament

Support wire

Casing contact point

Base contact point

# Engineering Design Process

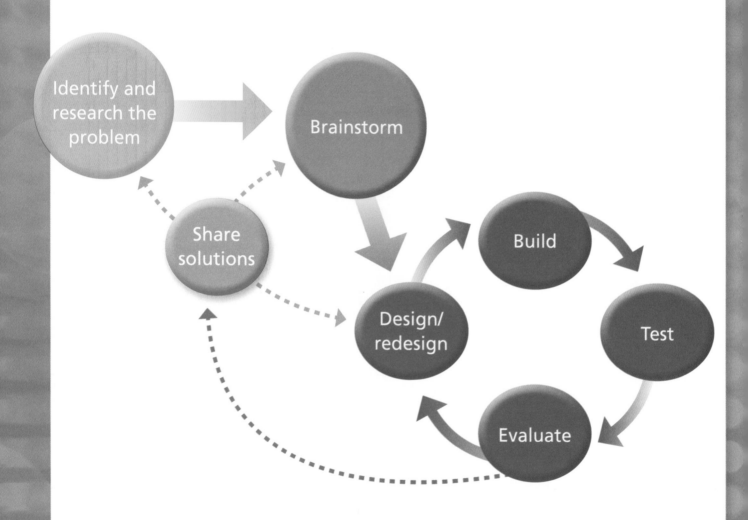

# Motor Dissection A

## Motors

A motor that runs on a D-cell is a direct-current motor. A direct-current motor has two main parts: permanent magnets and electromagnets.

A simple motor is like a tin can with two permanent magnets stuck inside. In the center of the can, there is a shaft that has two or more iron cores attached. A lot of wire is wrapped around each of the cores to make electromagnets.

### The Parts of a Simple Motor

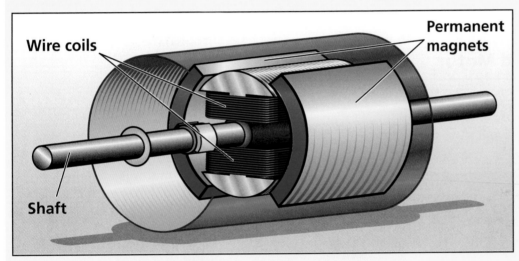

Wire coils

Permanent magnets

Shaft

Current from a D-cell flows through the wire coils, creating an electromagnetic field.

### The Magnets in a Simple Motor

Electromagnets

Permanent magnets

The electromagnet rotates between the poles of the fixed permanent magnets. This interaction spins the motor shaft.

# Motor Dissection B

Imagine taking the permanent magnets and shaft out of the can. Take off all but one of the wire coils. The simplified motor would look like number 1 below.

Connect the wire to a source of electric current, such as a D-cell. The flow of current makes a magnetic field around the wire (number 2).

When current flows in the coil, the coil becomes an electromagnet. The magnetic field of the electromagnet is repelled by the fields of the permanent magnets (like poles repel). This pushes the electromagnet away. The push causes the shaft to rotate.

But there is more to the design. When the shaft rotates, contact between the D-cell and the electromagnet is broken. The current stops flowing in the coil. The electromagnetism stops briefly. But momentum keeps the shaft rotating until contact is made again.

## Simplified Motor

**1**

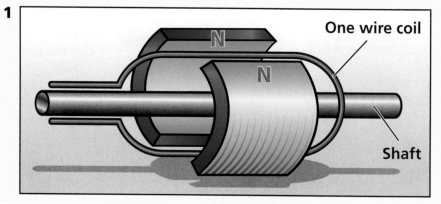

One wire coil

N

N

Shaft

This simplified motor is made up of two permanent magnets and one wire coil.

## Electromagnet

**2**

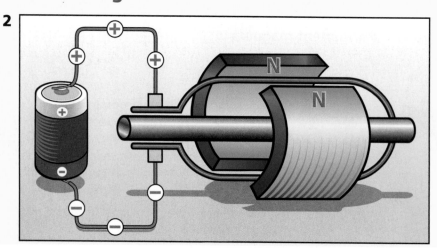

N

N

When the wire coil is connected to a D-cell, the device becomes an electromagnet.

As the shaft rotates a little farther, contact is made again. This creates the electromagnetic field in the coil again. The coil gets another magnetic push to keep the shaft turning.

The shaft gets hundreds of little magnetic pushes every second. The motor uses electric current and magnetism to produce motion.

## An Electric Motor

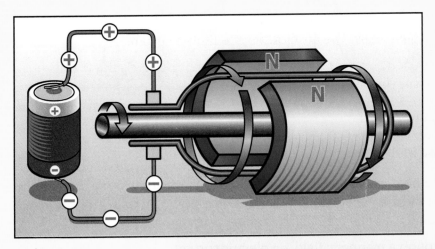

As the shaft rotates, contact between the D-cell and the coil is broken. As the shaft rotates a little farther, contact is made again.

Can you identify the shaft and wire coils in this electric motor?

# Generator Dissection

## How a Generator Works

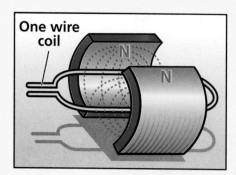

One wire coil

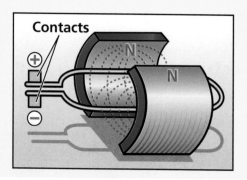

Contacts

When a wire coil turns in a magnetic field, an electric current is created in the wire.

Electricity flows from the generator when contact is made with the end of the wire coil.

There is a magnetic field between the poles of the permanent magnets. Set a shaft with wire coils between the magnets. Take off all but one of the wire coils. The simplified generator would look like the diagrams above.

When a wire passes through a magnetic field, an electric current is created in the wire. If you rotate a wire coil in a magnetic field, it will pass through the field hundreds of times in a second. This makes a continuous flow of electric current.

What turns the wire coil in the magnetic field? It could be many things. A windmill can be attached to the shaft. Wind can then rotate the coil. Water flowing downhill, steam from a boiler, or a gas engine can also be used to rotate the coil.

As a last resort, you can put a crank on the end of a generator shaft. A crank lets you turn the wire coil by hand to generate a little electricity.

## Hand-powered Generator

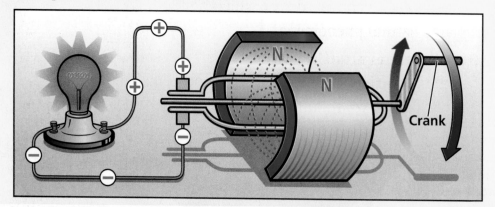

Crank

By turning the crank on the end of the shaft, you can generate electricity by hand.

# Science Practices

1. **Asking questions.** Scientists ask questions to guide their investigations. This helps them learn more about how the world works.

2. **Developing and using models.** Scientists develop models to represent how things work and to test their explanations.

3. **Planning and carrying out investigations.** Scientists plan and conduct investigations in the field and in laboratories. Their goal is to collect data that test their explanations.

4. **Analyzing and interpreting data.** Patterns and trends in data are not always obvious. Scientists make tables and graphs. They use statistical analysis to look for patterns.

5. **Using mathematics and computational thinking.** Scientists measure physical properties. They use computation and math to analyze data. They use mathematics to construct simulations, solve equations, and represent different variables.

6. **Constructing explanations.** Scientists construct explanations based on observations and data. An explanation becomes an accepted theory when there are many pieces of evidence to support it.

7. **Engaging in argument from evidence.** Scientists use argumentation to listen to, compare, and evaluate all possible explanations. Then they decide which best explains natural phenomena.

8. **Obtaining, evaluating, and communicating information.** Scientists must be able to communicate clearly. They must evaluate others' ideas. They must convince others to agree with their theories.

**Are you a scientist?**

# Engineering Practices

1. **Defining problems.** Engineers ask questions to make sure they understand problems they are trying to solve. They need to understand the constraints that are placed on their designs.

2. **Developing and using models.** Engineers develop and use models to represent systems they are designing. Then they test their models before building the actual object or structure.

3. **Planning and carrying out investigations.** Engineers plan and conduct investigations. They need to make sure that their designed systems are durable, effective, and efficient.

4. **Analyzing and interpreting data.** Engineers collect and analyze data when they test their designs. They compare different solutions. They use the data to make sure that they match the given criteria and constraints.

5. **Using mathematics and computational thinking.** Engineers measure physical properties. They use computation and math to analyze data. They use mathematics to construct simulations, solve equations, and represent different variables.

6. **Designing solutions.** Engineers find solutions. They propose solutions based on desired function, cost, safety, how good it looks, and meeting legal requirements.

7. **Engaging in argument from evidence.** Engineers use argumentation to listen to, compare, and evaluate all possible ideas and methods to solve a problem.

8. **Obtaining, evaluating, and communicating information.** Engineers must be able to communicate clearly. They must evaluate other's ideas. They must convince others of the merits of their designs.

**Are you an engineer?**

# Science Safety Rules

1. Always follow the safety procedures outlined by your teacher. Follow directions, and ask questions if you're unsure of what to do.

2. Never put any material in your mouth. Do not taste any material or chemical unless your teacher specifically tells you to do so.

3. Do not smell any unknown material. If your teacher asks you to smell a material, wave a hand over it to bring the scent toward your nose.

4. Avoid touching your face, mouth, ears, eyes, or nose while working with chemicals, plants, or animals. Tell your teacher if you have any allergies.

5. Always wash your hands with soap and warm water immediately after using chemicals (including common chemicals, such as salt and dyes) and handling natural materials or organisms.

6. Do not mix unknown chemicals just to see what might happen.

7. Always wear safety goggles when working with liquids, chemicals, and sharp or pointed tools. Tell your teacher if you wear contact lenses.

8. Clean up spills immediately. Report all spills, accidents, and injuries to your teacher.

9. Treat animals with respect, caution, and consideration.

10. Never use the mirror of a microscope to reflect direct sunlight. The bright light can cause permanent eye damage.

# Glossary

**acceleration** a change to an object's speed

**attract** to pull toward each other

**automobile** a wheeled vehicle propelled by a motor

**battery** a source of stored chemical energy

**brush** a conductor that makes contact inside a motor or a generator

**circuit** a pathway for the flow of electricity

**climate change** the change of worldwide average temperature and weather conditions

**closed circuit** a complete circuit through which electricity flows

**commutator** a switching mechanism on a motor or generator

**compass** an instrument that uses a free-rotating magnetic needle to show direction

**complete circuit** a circuit with all the necessary connections for electricity to flow

**component** one item in a circuit

**compress** to squeeze or press

**conductor** a material capable of transmitting energy, particularly in the forms of heat and electricity

**constraint** a restriction or limitation

**contact point** the place on a component where connections are made to allow electricity to flow

**core** in an electromagnet, the material around which a coil of insulated wire is wound

**criterion** (plural: criteria) a standard for evaluating or testing something

**drag** the resistance of air to objects moving through it

**electric current** the flow of electricity through a conductor

**electric force** the force between two charged objects

**electromagnet** a piece of iron that becomes a temporary magnet when electricity flows through coils of insulated wire wrapped around it

**electromagnetic force** a fundamental force of nature that acts between charged particles and creates electric and magnetic fields

**electromagnetism** the interaction of electric and magnetic fields

**electron** a tiny part of an atom that has a negative charge and moves around the nucleus

**energy** the ability to apply force over a distance

**engineer** a designer who tests ways to accomplish a goal or solve a problem

**filament** the material in a lightbulb (usually a thin wire) that makes light when heated by an electric current

**force** a push or a pull

**fossil fuel** the remains of organisms that lived long ago preserved as oil, coal, or natural gas

**friction** a force acting between surfaces in contact. Friction acts to resist motion.

**fuel** a material that stores energy

**generator** a device that produces electricity from motion

**gravitational field** an invisible area of gravitational influence around a mass

**gravity** the natural attraction between masses. On Earth, all objects are pulled toward the center of Earth.

**greenhouse gas** a gas that absorbs and radiates energy in the atmosphere, effectively trapping heat

**incandescent lightbulb** a device that gives off light and heat when electric current runs through a filament

**incomplete circuit** a circuit that has a break in it

**induced magnetism** the influence of a magnetic field on iron, which makes the iron a temporary magnet

**insulator** a material that prevents the flow of electricity, commonly plastic, rubber, glass, or air

**interaction** to act on and be acted upon by one or more objects

**lamp** a device that produces light, either by electricity or burning oil, gas, or wax

**maglev** a device utilizing magnetic levitation

**magnet** an object that responds to magnetic force

**magnetic field** an area of magnetic influence around a magnet

**magnetism** a property of certain kinds of materials that causes them to attract iron or steel

**motor** a device that produces motion from electricity

**net force** the sum of all the forces acting on a mass

**newton (N)** the standard unit for measuring force in the metric system

**nonrenewable** a material that cannot be replaced once used up

**open circuit** an incomplete circuit through which electricity will not flow

**particle** a very small piece or part

**permanent magnet** an object that sticks to iron

**pole** the end of a magnet

**potential energy** energy that matter has because of its position

**power grid** a complex network of wires that delivers electricity to homes and other buildings

**renewable** able to be replaced or restored by nature

**repel** to push away from each other

**semiconductor** a material, such as silicon, that has less electric conductivity than a conductor but more than an insulator

**shaft** the part of a motor that rotates when energy is transferred to it

**solar cell** a silicon structure that converts sunlight into electrical energy and is used as a power source

**spring scale** an instrument used to measure force

**static** not moving

**sustainable** able to be used without being completely used up or destroyed

**temporary magnet** a piece of iron that behaves like a magnet only when it is under the influence of an external magnetic field

**turbine** a generator powered by the flow of air, steam, or another fluid

**weight** the force of gravity on a mass

# Index

**A**
**acceleration**, 13, 76
**Aristotle**, 9, 11, 14
**atom**, 32, 36
**attract**, 21, 76
**automobile**, 48, 76

**B**
**battery**, 25–28, 36–39, 43, 48–51, 53–55, 61, 76
**Bell, Alexander Graham**, 43
**brush**, 58, 76

**C**
**chemical energy**, 51
**chemical reaction**, 36, 54, 57
**circuit**, 26, 39–41, 45, 76
**circuitry**, 25–30, 36
**climate change**, 30, 54, 57, 62, 76
**closed circuit**, 26, 76
**communication**, 42–43
**commutator**, 58, 76
**compact fluorescent lamp**, 29
**compass**, 6–7, 76
**complete circuit**, 25–27, 36, 76
**component**, 26, 76
**compress**, 6–7, 76
**conductor**, 26, 76
**constraint**, 50–52, 76
**contact point**, 26, 76
**core**, 41, 45, 76
**criterion**, 50, 76

**D**
**Davidson, Robert**, 49
**drag**, 46, 76

**E**
**Edison, Thomas**, 28, 29
**electric car**, 47–55
**electric current**, 25, 36, 38–41, 44–46, 51, 76
**electric force**, 32, 76
**electric signal**, 44
**electricity**, 27–28, 31–37, 57, 60–61
**electromagnet**, 39 43, 44–46, 76
**electromagnetic engineering**, 42–46
**electromagnetic field**, 43
**electromagnetic force**, 41, 76
**electromagnetic train**, 42, 46
**electromagnetism**, 38–41, 76
**electron**, 32, 36, 76
**energy**, 25, 29, 37, 44, 48, 50, 56-62, 76
**engineer/engineering**, 30, 42–46, 48, 60, 62, 76
**engineering practices**, 74

**F**
**Faraday, Michael**, 41
**filament**, 27, 28, 76
**fluorescent lamp**, 29
**force**, 3–7, 42, 76

**fossil fuel**, 54, 57, 58, 59, 76
**friction**, 8–14, 46, 52, 76
**fuel**, 54, 76

**G**
**Galilei, Galileo**, 10–11, 13, 14
**generator**, 52, 58, 60, 76
**global warming**, 57, 59
**gravitational field**, 20, 77
**gravity**, 5, 11, 12, 77
**greenhouse gas**, 57, 59, 60, 77

**H**
**heat**, 28, 29, 44
**hydroelectric power**, 60

**I**
**incandescent bulb**, 28, 29, 77
**incomplete circuit**, 26, 77
**induced magnetism**, 21, 77
**induction heating**, 44
**insulator**, 34, 77
**interaction**, 4, 77

**L**
**lamp**, 27, 77
**Latimer, Lewis**, 28
**light**, 28, 29
**lightbulb**, 25, 27–30
**light-emitting diode (LED)**, 30

**M**
**maglev train**, 46, 77
**magnet**, 19–23, 46, 77
**magnetic field**, 20, 23–24, 39, 46, 77
**magnetic force**, 19–24
**magnetic resonance imaging (MRI)**, 46
**magnetism**, 19, 22, 45, 77
**microphone**, 44
**Morse, Samuel F. B.**, 43
**motion**, 13, 18
**motor**, 44, 77

**N**
**net force**, 15–18, 77
**Newton, Isaac**, 12–14
**newton (N)**, 6–7, 16, 77
**nonrenewable**, 59, 77
**nuclear power**, 59

**O**
**open circuit**, 26, 77
**Ørsted, Hans Christian**, 39, 41

**P**
**particle**, 22, 23–24, 77
**permanent magnet**, 20, 22, 43, 58, 77
**photovoltaic technology**, 61
**Planté, Gaston**, 49
**pole**, 20–22, 77
**potential energy**, 61, 77
**power grid**, 57, 58, 77

**R**
**renewable**, 54, 62, 77
**repel**, 21, 77

**S**
**science practices**, 73
**science safety rules**, 75
**semiconductor**, 34, 77
**series circuit**, 45
**shaft**, 6, 58, 77
**solar cell**, 61, 77
**solar energy**, 54, 61
**sound wave**, 43–44
**spring scale**, 6–7, 77
**static**, 31, 33, 77
**sustainable**, 53, 54, 55, 56, 77

**T**
**telegraph**, 43
**telephone**, 43
**temporary magnet**, 21, 24, 41, 77
**thermal energy**, 53, 57
**turbine**, 58–61, 77

**V**
**Volta, Alessandro**, 38

**W**
**weight**, 6–7, 77
**wind energy**, 54, 61